Modular Functions in Analytic Number Theory

MARVIN I. KNOPP
University of Illinois at Chicago

MARKHAM PUBLISHING COMPANY
Chicago

MARKHAM MATHEMATICS SERIES
William J. LeVeque, Editor

Anderson, *Graph Theory and Finite Combinatorics*
Cantrell and Edwards, eds., *Topology of Manifolds*
Gaal, *Galois Theory of Polynomials with Examples*
Gioia, *The Theory of Numbers: An Introduction*
Greenspan, *Introduction to Numerical Analysis and Applications*
Hu, *Calculus*
Hu, *Cohomology Theory*
Hu, *Elementary Functions and Coordinate Geometry*
Hu, *Linear Algebra*
Knopp, *Theory of Area*
Knopp, *Modular Functions in Analytic Number Theory*
Peterson, *Foundations of Algebra and Number Theory*
Stark, *An Introduction to Number Theory*

LECTURES IN ADVANCED MATHEMATICS

Davenport, *1. Multiplicative Number Theory*
Storer, *2. Cyclotomy and Difference Sets*
Engeler, *3. Formal Languages: Automata and Structures*
Garsia, *4. Topics in Almost Everywhere Convergence*

Printed in the United States of America
Library of Congress Catalog Card Number: 79–122307
Standard Book Number: 8410–1000–5

To My Wife and Children

PREFACE

An accurate (though uninspiring) title for this book would have been *Applications of the Theory of the Modular Forms* $\eta(\tau)$ *and* $\vartheta(\tau)$ *to the Number-Theoretic functions* $p(n)$ *and* $r_s(n)$, *respectively.* This is accurate because, except in the first two chapters, we deal exclusively with these two modular forms and these two number-theoretic functions (see Chapter 3 for definitions). However, at the heart of these particular applications to the treatment of these specific number-theoretic functions lies the general theory of automorphic functions, a theory of far-reaching significance with important connections to a great many fields of mathematics. Indeed, together with Riemann surface theory, analytic number theory has provided the principal impetus for the development over the last century of the theory of automorphic functions.

Chapter 1 is an introduction to the modular group $\Gamma(1)$ and certain of its congruence subgroups. Of primary importance here is the construction of the familiar fundamental region for $\Gamma(1)$ (Theorem 1), and the construction of what I have chosen to call a "standard fundamental region" for an arbitrary subgroup of finite index in $\Gamma(1)$ (Theorem 12). In this respect the restriction to subgroups of finite index in the modular group affords an important simplification in comparison to the theory of general discontinuous groups, in which the discussion of the fundamental region and its boundary amounts to a major undertaking.

Chapter 2 gives an introduction to the theory of modular functions and forms for subgroups of finite index in $\Gamma(1)$ which, although brief, is sufficient for the applications we give later. From the point of view of the later applications, the most important results of this chapter are Theorem 7, which states that a *bounded* modular function is constant, and Theorem 10, which gives an asymptotic estimate for the Fourier coefficients of a cusp form. These results, as well as the others of the chapter, hold for a wide class of discontinuous groups with no essential change in the proofs required. Chapters 1 and 2 alone might well serve those readers who are interested in an introduction to the theory of modular functions and forms, but who have no particular interest in the applications to number theory.

In Chapter 3 we introduce the number-theoretic functions $p(n)$ and $r_s(n)$ and their respective generating functions $\eta(\tau)$ and $\vartheta^s(\tau)$, and carry out a fairly extensive study of $\eta(\tau)$ and $\vartheta(\tau)$. Our main purpose here is to derive the transformation properties of $\eta(\tau)$ and $\vartheta(\tau)$ and show that these functions are in fact modular forms (Theorems 10 and 13). We deal first with the transformation properties of $\eta(\tau)$ under $\Gamma(1)$, and then derive the corresponding properties of $\vartheta(\tau)$ under the subgroup Γ_ϑ (of index 3) by making use of Theorem 12, which expresses $\vartheta(\tau)$ in terms of $\eta(\tau)$. This link between $\vartheta(\tau)$ and $\eta(\tau)$ follows from the well-known product representation of $\vartheta(\tau)$ (Theorem 11), which, in turn, is a consequence of the famous Jacobi identity (Theorem 3). We also use Theorem 3 to derive the identity of Euler (Corollary 5) and a second identity due to Jacobi (Corollary 6). Our proof of the transformation properties of $\eta(\tau)$ is based upon the Poisson sum formula (Theorem 7), which is also applied in Chapter 5 in our proof of the Lipschitz summation formula. In Chapter 6 we state and prove exact formulas for the multiplier systems of $\eta(\tau)$ and $\vartheta(\tau)$. Chapters 3 and 4 can be read independently of the rest of the book.

Chapter 5 is devoted to the function $r_s(n)$. The main results are Theorem 2, which expresses $r_s(n)$ asymptotically in terms of the *singular series* $\rho_s(n)$, for each integer $s \geqslant 5$, and Theorem 3, which shows that the asymptotic formula of Theorem 2 is an *exact* formula for $5 \leqslant s \leqslant 8$. Broadly speaking, our method is to prove that the Eisenstein series $\psi_s(\tau)$ is a modular form with exactly the same transformation properties as $\vartheta^s(\tau)$, and then to compare $\psi_s(\tau)$ with $\vartheta^s(\tau)$. The Lipschitz summation formula (Theorem 4) is used to show that $\rho_s(n)$ is the nth Fourier coefficient of $\psi_s(\tau)$ (Theorem 9), and the results of Chapter 2 are ultimately applied to prove Theorems 2 and 3. The construction of the Eisenstein series is a quite general principle which is by no means restricted to the particular case of $\psi_s(\tau)$; but, for our purposes, we find it sufficient to restrict ourselves to this case. Chapter 5 closes with proofs that $\rho_s(n)$ has order of growth $n^{s/2-1}$ (Theorem 14) and that for $s > 8$, $\psi_s(\tau)$ is no longer equal to $\vartheta^s(\tau)$ (Theorem 17).

Chapter 6 begins the study of the properties of $p(n)$. In this chapter the "circle method" is applied to give a good asymptotic formula for $p(n)$ (Theorem 2). This asymptotic formula is closely related to Rademacher's well-known exact formula which we give, without proof, at the beginning of the chapter. Chapter 6 requires as preparation only the results on $\eta(\tau)$ contained in Chapter 3.

The high point of the book, in my opinion, is the complete account in Chapters 7 and 8 of the celebrated Ramanujan congruences for $p(n)$ modulo all powers of 5 and all powers of 7. I am deeply indebted to A. O. L. Atkin, who made available to me an unpublished manuscript in which he presents a significant and highly readable simplification of Watson's proof of the Ramanujan congruences modulo powers of 5 and 7. Without this access to Atkin's work, the inclusion here of the proofs of these congruences would not have been possible. In fact, Chapter 8 is essentially an expanded version of the first half of the Atkin manuscript, with even the notation retained. Chapter 7 contains the modular function theory required as background for Chapter 8. Most important here and essential in the proof of the Ramanujan congruences are the "modular equations" for the primes 5 and 7, which are stated and proved in Sections 6 and 7 of Chapter 7. The material we develop for the proof of the modular equations also enables us to present, in Section 4 of the chapter, a proof of the Ramanujan congruence modulo 11, from Morris Newman. Chapters 7 and 8 can be read immediately after Chapter 4. The reader already familiar with the modular equations (or willing to accept them on faith) can study Chapter 8 independently of the rest of the book.

I have tried to keep the book self-contained for those readers who have had a good first-year graduate course in analysis; and, in particular, I have assumed readers to be familiar with the Cauchy theory and the Lebesgue theorem of dominated convergence. Where material has been quoted that would probably not appear in such a course, a specific reference is given. An example of this occurs in Chapter 3 in the proof of the Poisson sum formula. There we apply Fejer's theorem on Fourier series and supply a reference to Titchmarsh. Previous experience with the Jacobi symbol of number theory would be very useful, although not essential.

Most of the material presented here has been developed from various lecture courses I have given in analytic number theory at the University of Wisconsin in Madison. This book is intended to begin to fill what I consider a rather major gap in the literature of number theory. This is not to say that I claim originality for any portion of the contents; in fact, only the arrangement is my own, the newer material for the most part having been gleaned from various sources in the research journals. (An exception is Chapter 8, which, as mentioned above, is taken from the unpublished manuscript of Atkin. However, there is no other available book that brings together all of the material necessary for a reasonably good discussion of the functions $p(n)$ and

$r_s(n)$ from the point of view of modular functions. As a result the student or mature mathematician wishing to inform himself on the subject has had the difficult and time-consuming task of collecting the necessary material from the various sources. It is my hope and belief that the publication of this book will ease the situation sufficiently so that many more mathematicians will find a serious and lasting interest in what I believe to be a truly important and beautiful part of mathematics.

It is a pleasure to thank Emil Grosswald and Bruce Berndt, who carefully read large portions of the manuscript, suggested many improvements, and uncovered errors. My thanks are also due to Phil Gerould of Markham, who helped bring this book to fruition, and to my wife, Josephine, who did an excellent job of typing the manuscript without the benefit of a mathematical typewriter.

I am especially indebted to Paul Bateman, my teacher, who first introduced me to the subject of this book. It was through Professor Bateman that I was first exposed to the thinking of the late Professor Hans Rademacher, who has had a profound influence upon all of us who work in this area of mathematics.

M.I.K.

Madison, Wisconsin
May, 1970

CONTENTS

Chapter 1

THE MODULAR GROUP AND CERTAIN SUBGROUPS

1. THE MODULAR GROUP

The *modular group*, denoted $\Gamma(1)$, is the set of linear fractional transformations V, $V\tau = (a\tau + b)/(c\tau + d)$, where a, b, c, d are rational integers such that $ad - bc = 1$. Here τ is a complex variable. It is easy to verify that $\Gamma(1)$ is indeed a group and that each element of $\Gamma(1)$ preserves the upper half-plane, the real line, and the lower half-plane. Throughout this book we shall use $\mathscr{H}$ to denote the upper half-plane and Z for the set of rational integers.

We can interpret $\Gamma(1)$ as the group of 2 by 2 rational integral matrices of determinant one, if we keep in mind that the matrices $\begin{pmatrix} a & b \\ c & d \end{pmatrix}$ and $\begin{pmatrix} -a & -b \\ -c & -d \end{pmatrix}$ are to be identified. [Expressed more technically, $\Gamma(1)$ is isomorphic to the matrix group modulo its center. The center consists in this case of $\pm\begin{pmatrix} 1 & 0 \\ 0 & 1 \end{pmatrix}$.] Composition of linear fractional transformations corresponds to matrix multiplication. For let

$$V_i = \begin{pmatrix} a_i & b_i \\ c_i & d_i \end{pmatrix}, \quad V_i\tau = \frac{a_i\tau + b_i}{c_i\tau + d_i}; \qquad i = 1, 2.$$

Then

$$V_1V_2 = \begin{pmatrix} a_1a_2 + b_1c_2 & a_1b_2 + b_1d_2 \\ c_1a_2 + d_1c_2 & c_1b_2 + d_1d_2 \end{pmatrix},$$

where V_1 and V_2 are considered as matrices, while

$$V_1(V_2\tau) = \frac{a_1V_2\tau + b_1}{c_1V_2\tau + d_2}$$
$$= \frac{(a_1a_2 + b_1c_2)\tau + a_1b_2 + b_1d_2}{(c_1a_2 + d_1c_2)\tau + c_1b_2 + d_1d_2},$$

where V_1 and V_2 are considered as linear fractional transformations. We shall frequently use this correspondence to simplify our calculations.

Suppose that Γ is a subgroup of $\Gamma(1)$. We say that the complex numbers τ_1 and τ_2 are *equivalent with respect to* Γ if there exists $V \in \Gamma$ such that $V\tau_1 = \tau_2$.

Definition. Let Γ be a subgroup of $\Gamma(1)$. A *fundamental region* (F.R.) *for* Γ is an open subset $\mathscr{R}$ of $\mathscr{H}$ such that

(a) no two *distinct* points of $\mathscr{R}$ are equivalent with respect to Γ, and

(b) every point of $\mathscr{H}$ is equivalent to some point of $\bar{\mathscr{R}}$, the closure of $\mathscr{R}$. (The "closure" is here meant in the sense of the ordinary topology of the Riemann sphere.)

Remark. The fundamental region as defined here is far from unique. Any closed set without an interior can be removed from $\mathscr{R}$ and we still obtain a fundamental region. For example, any closed line segment can be removed from $\mathscr{R}$. Furthermore, a piece of $\mathscr{R}$ can be removed and transformed by an element of Γ. The interior of the resulting set is again a fundamental region. Notice that a fundamental region *need not be a region*; that is, it need not be connected.

We do not here consider the question of the existence of a fundamental region for an arbitrary subgroup Γ. In this book we deal only with subgroups of finite index in $\Gamma(1)$ and for these we shall explicitly construct the fundamental regions.

We now give two simple examples to illustrate the notion of a fundamental region and also to indicate the kinds of nonuniqueness that can occur.

Examples. 1. Let S be the linear fractional transformation given by $S\tau = \tau + 1$. We may also write S in the matrix form

$$S = \begin{pmatrix} 1 & 1 \\ 0 & 1 \end{pmatrix}.$$

Let Γ be the subgroup of $\Gamma(1)$ generated by S. Then Γ consists of all translations $S^n\tau = \tau + n$, $n \in Z$. An obvious choice for a F.R. is the set given by the inequalities $\operatorname{Im}\tau > 0$, $|\operatorname{Re}\tau| < \frac{1}{2}$. Another choice is that set given by $\operatorname{Im}\tau > 0$, $0 < \operatorname{Re}\tau < 1$. Further choices follow:

$$\{\tau \in \mathscr{H} | 0 < \operatorname{Re}\tau < \tfrac{1}{2}\} \cup \{\tau \in \mathscr{H} | \tfrac{1}{2} < \operatorname{Re}\tau < 1\},$$

$$\{\tau \in \mathscr{H} | 0 < \operatorname{Re}\tau < \tfrac{1}{2}\} \cup \{\tau \in \mathscr{H} | \tfrac{11}{2} < \operatorname{Re}\tau < 6\}.$$

2. Let T be the linear fractional transformation $T\tau = -1/\tau$. In matrix form we write $T = \begin{pmatrix} 0 & -1 \\ 1 & 0 \end{pmatrix}$. As a linear fractional transformation, T has order 2 and therefore generates a subgroup of order 2 in $\Gamma(1)$. As a F.R. for this subgroup we may choose $\{\tau \in \mathscr{H} \,|\, |\tau| < 1\}$. Alternatively, $\{\tau \in \mathscr{H} \,|\, |\tau| > 1\}$ will serve. The reader should be able to verify that $\{\tau \in \mathscr{H} \,|\, \operatorname{Re}\tau > 0\}$ is also a F.R. for this group.

Incidentally, we shall show in Corollary 3 that S and T together generate all of $\Gamma(1)$.

2. A FUNDAMENTAL REGION FOR $\Gamma(1)$

Theorem 1. As a F.R. for $\Gamma(1)$ we may choose the region

$$\mathscr{R}(\Gamma(1)) = \{\tau \in \mathscr{H} \,|\, |\tau| > 1, |\operatorname{Re}\tau| < \tfrac{1}{2}\}.$$

Proof. Let

$$\mathscr{R}^* = \{\tau \in \mathscr{H} \,|\, |\operatorname{Re}\tau| < \tfrac{1}{2}, |c\tau + d| > 1 \quad \text{for all } c,d \in Z \text{ with } (c,d) = 1 \text{ and } c \neq 0\}.$$

We shall first show that $\mathscr{R}^* = \mathscr{R}(\Gamma(1))$. We put $\mathscr{R} = \mathscr{R}(\Gamma(1))$ for the remainder of the proof. Obviously $\mathscr{R}^* \subset \mathscr{R}$, as can be seen by choosing $c = 1$, $d = 0$ in the defining conditions for $\mathscr{R}^*$.

To show the reverse inclusion, let $\tau = x + iy \in \mathscr{R}$ and let $(c,d) = 1$, with $c \neq 0$. Then

$$\begin{aligned} |c\tau + d|^2 &= c^2(x^2 + y^2) + 2cdx + d^2 \\ &> c^2 - |cd| + d^2 = (|c| - |d|)^2 + |cd| \geqslant 1, \end{aligned}$$

so that $\tau \in \mathscr{R}^*$, and it follows that $\mathscr{R}^* \supset \mathscr{R}$. Hence $\mathscr{R}^* = \mathscr{R}$. It remains to show that $\mathscr{R}^*$ is a F.R. for $\Gamma(1)$.

First we observe that

$$\operatorname{Im}\left(\frac{a\tau + b}{c\tau + d}\right) = \frac{\operatorname{Im}\tau}{|c\tau + d|^2}$$

as long as $ad - bc = 1$. Thus

$$\operatorname{Im}\left(\frac{a\tau + b}{c\tau + d}\right) \geqslant \operatorname{Im}\tau$$

if and only if $|c\tau + d|^2 \leqslant 1$; that is, $(cx + d)^2 + c^2y^2 \leqslant 1$. With fixed x and y this latter inequality has only finitely many solutions in integers c, d. We want to show first that if $\tau_0 \in \mathscr{H}$, then τ_0 is equivalent with respect to $\Gamma(1)$ to a point in $\bar{\mathscr{R}}^* = \bar{\mathscr{R}}$. We have already shown that either $|c\tau_0 + d| > 1$ for *all* pairs c, d with $(c,d) = 1$, $c \neq 0$, or $|c\tau_0 + d| \leqslant 1$ for finitely many pairs c, d with $(c,d) = 1, c \neq 0$. In the first case we simply choose an integer m such that $|\operatorname{Re}(\tau_0 + m)| \leqslant \frac{1}{2}$; it is still the case that $|c(\tau_0 + m) + d| > 1$ for all pairs c, d with $(c,d) = 1$, $c \neq 0$. Thus $\tau_0 + m = (\tau_0 + m)/(0\tau_0 + 1) \in \bar{\mathscr{R}}^*$ in this case.

In the second situation choose the pair c', d', with $(c',d') = 1$, which *minimizes* $|c\tau_0 + d|$. Since $(c', d') = 1$ we can determine integers a', b' with $a'd' - b'c' = 1$; that is,

$$\begin{pmatrix} a'b' \\ c'd' \end{pmatrix} \in \Gamma(1).$$

Then $\omega_0 = (a'\tau_0 + b')/(c'\tau_0 + d')$ and τ_0 are equivalent with respect to $\Gamma(1)$, and

$$(1) \qquad |c\omega_0 + d| = \left|\frac{(a'c + dc')\tau_0 + (b'c + dd')}{c'\tau_0 + d'}\right|.$$

We claim that $|c\omega_0 + d| \geqslant 1$ for all pairs c, d with $(c,d) = 1$. For choose a, b such that $ad - bc = 1$. Then

$$\begin{aligned}(aa' + bc')(b'c + dd') - (ab' + bd')(a'c + dc')\\ = (ad - bc)(a'd' - b'c') = 1,\end{aligned}$$

and therefore $(b'c + dd', a'c + dc') = 1$. Hence

$$|(a'c + dc')\tau_0 + (b'c + dd')| \geqslant |c'\tau_0 + d'|,$$

by the minimality of the right-hand side. Thus $|c\omega_0 + d| \geqslant 1$ follows from (1). If ω_0 does not satisfy $|\operatorname{Re}\omega_0| \leqslant \frac{1}{2}$, we replace ω_0 by $\omega_0 + m$ for a suitably chosen integer m, to obtain $\omega_0 + m \in \bar{\mathscr{R}}^*$. Note that $\omega_0 + m$ is equivalent to τ_0 with respect to $\Gamma(1)$.

We must yet prove that no two points of $\mathscr{R}^*$ are equivalent with respect to $\Gamma(1)$. Suppose $\tau \in \mathscr{R}^*$; then $|c\tau + d| > 1$ for all pairs c, d with $(c,d) = 1$,

$c \neq 0$. Let

$$\omega = \frac{a\tau + b}{c\tau + d}, \qquad c \neq 0, \qquad \begin{pmatrix} a & b \\ c & d \end{pmatrix} \in \Gamma(1).$$

Then $|-c\omega + a| = |c\tau + d|^{-1} < 1$. Also $(-c, a) = 1$, in view of the fact that $ad - bc = 1$. Thus $\omega \notin \mathscr{R}^*$, by definition of $\mathscr{R}^*$. On the other hand, if $c = 0$, that is, if $\omega = \tau + b$, then $|\mathrm{Re}\,\omega| > \frac{1}{2}$ unless $b = 0$. Thus a point ω equivalent to $\tau \in \mathscr{R}^*$ with respect to $\Gamma(1)$ cannot be in $\mathscr{R}^*$, unless $\tau = \omega$. This completes the proof.

3. SOME SUBGROUPS OF $\Gamma(1)$

Definition. Let n be a positive integer. Define *the principal congruence subgroup of level n*, $\Gamma(n)$, by $a \equiv d \equiv \pm 1$, $b \equiv c \equiv 0 \pmod n$. In the matrix notation we have, given $M \in \Gamma(1)$, that $M \in \Gamma(n)$, provided $M \equiv \pm I \pmod n$, where $I = \begin{pmatrix} 1 & 0 \\ 0 & 1 \end{pmatrix}$, and the congruence is elementwise. Let $\Gamma_0(n)$ be the subgroup of $\Gamma(1)$ defined by $c \equiv 0 \pmod n$ and let $\Gamma^0(n)$ be the subgroup defined by $b \equiv 0 \pmod n$. Let Γ_ϑ be that subgroup of $\Gamma(1)$ generated by S^2, T.

It is easy to see that $\Gamma(n)$ is normal in $\Gamma(1)$, while $\Gamma_0(n)$, $\Gamma^0(n)$, Γ_ϑ are not. In fact, $T\Gamma_0(n)T^{-1} = \Gamma^0(n)$, while $S\Gamma_\vartheta S^{-1} = \Gamma^0(2)$ (see Lemma 8, below). We shall presently obtain coset decompositions of $\Gamma(1)$ modulo these various groups.

Theorem 2. Let n be an integer with $1 \leqslant n \leqslant 4$ and let Γ_1 be the group generated by $\tau' = \tau + \sqrt{n}$ and $\tau' = -1/\tau$ $\left[\text{in matrix notation, by } \begin{pmatrix} 1 & \sqrt{n} \\ 0 & 1 \end{pmatrix}\right.$ and $\left. T = \begin{pmatrix} 0 & -1 \\ 1 & 0 \end{pmatrix}\right]$. Let Γ_2 be the set of all linear fractional transformations of the two types

(a) $$\tau' = \frac{a\tau + b\sqrt{n}}{c\sqrt{n}\tau + d}; \qquad a, b, c, d \in Z, \quad ad - nbc = 1,$$

(b) $$\tau' = \frac{a\sqrt{n}\tau + b}{c\tau + d\sqrt{n}}; \qquad a, b, c, d \in Z, \quad nad - bc = 1,$$

where Z is the set of all rational integers. Then Γ_2 is a group and $\Gamma_1 = \Gamma_2$.

Proof. It is easy to check that Γ_2 is a group. Since the generators of Γ_1 are in Γ_2, we have $\Gamma_1 \subset \Gamma_2$. We shall show that $\Gamma_2 \subset \Gamma_1$.

Let $\begin{pmatrix} a & b\sqrt{n} \\ c\sqrt{n} & d \end{pmatrix} \in \Gamma_2$. If $a = 0$, then $n = 1$ and the matrix is

$$\pm\begin{pmatrix} 0 & -1 \\ 1 & d \end{pmatrix} = \pm\begin{pmatrix} 0 & -1 \\ 1 & 0 \end{pmatrix}\begin{pmatrix} 1 & d \\ 0 & 1 \end{pmatrix} \in \Gamma_1.$$

If $b = 0$, then $a = d = \pm 1$, and the matrix is

$$\pm\begin{pmatrix} 1 & 0 \\ c\sqrt{n} & 1 \end{pmatrix} = \mp\begin{pmatrix} 0 & -1 \\ 1 & 0 \end{pmatrix}\begin{pmatrix} 1 & -c\sqrt{n} \\ 0 & 1 \end{pmatrix}\begin{pmatrix} 0 & -1 \\ 1 & 0 \end{pmatrix} \in \Gamma_1.$$

Hence we may assume that $a \neq 0$ and $b \neq 0$. If $t \in Z$, then

$$\begin{pmatrix} a & b\sqrt{n} \\ c\sqrt{n} & d \end{pmatrix}\begin{pmatrix} 1 & t\sqrt{n} \\ 0 & 1 \end{pmatrix} = \begin{pmatrix} a & b'\sqrt{n} \\ c\sqrt{n} & d' \end{pmatrix},$$

where $b' = at + b$ and $d' = ctn + d$. We claim we can choose $t \in Z$ so that $|b'\sqrt{n}| < |a|$. The condition $|b'\sqrt{n}| < |a|$ is equivalent to $-|a| < at\sqrt{n} + b\sqrt{n} < |a|$, or

(2) $$|a|(\sqrt{n}t - 1) < b\sqrt{n} < |a|(\sqrt{n}t + 1),$$

where we have replaced t by $-t$ if $a > 0$. We are thus claiming that we can always choose $t \in Z$ so that (2) is fulfilled.

If $1 \leqslant n \leqslant 4$, since $a \neq 0$, the closed intervals $[|a|(\sqrt{n}t - 1, |a|(\sqrt{n}t + 1)]$ with $t \in Z$, cover the entire real line. (This is not true if $n \geqslant 5$.) In fact, if $1 \leqslant n \leqslant 3$, the corresponding *open* intervals cover the entire real line. Thus (2) can be achieved at least in the cases $1 \leqslant n \leqslant 3$, while in the case $n = 4$ we can achieve (2) with the strict inequality replaced by

(3) $$|a|(\sqrt{n}t - 1) \leqslant b\sqrt{n} \leqslant |a|(\sqrt{n}t + 1).$$

However, if $n = 4$, then a is odd, so that an equality in (3) would be between integers of opposite parity. Thus we can achieve (2) also in this case.

Let $t \in Z$ be chosen so that (2) holds. If it happens that $b' = 0$, then $\begin{pmatrix} a & b'\sqrt{n} \\ c\sqrt{n} & d' \end{pmatrix} \in \Gamma_1$, as before, so that $\begin{pmatrix} a & b\sqrt{n} \\ c\sqrt{n} & d \end{pmatrix} \in \Gamma_1$. If $b' \neq 0$, let $q \in Z$ and consider

$$\begin{pmatrix} 0 & -1 \\ 1 & 0 \end{pmatrix}\begin{pmatrix} a & b'\sqrt{n} \\ c\sqrt{n} & d' \end{pmatrix}\begin{pmatrix} 0 & -1 \\ 1 & 0 \end{pmatrix}\begin{pmatrix} 1 & q\sqrt{n} \\ 0 & 1 \end{pmatrix} = \begin{pmatrix} -d' & c'\sqrt{n} \\ b'\sqrt{n} & -a' \end{pmatrix},$$

where $c' = c - d'q$ and $a' = a - nb'q$. In complete analogy with the proof of (2) we can prove that there exists $q \in Z$ such that

(4) $$|b'|\sqrt{n}(\sqrt{n}q - 1) < a < |b'|\sqrt{n}(\sqrt{n}q + 1).$$

That is, there exists $q \in Z$ such that $|a'| < |b'|\sqrt{n}$, provided $b' \neq 0$. Thus we have

$$\begin{pmatrix} 0 & -1 \\ 1 & 0 \end{pmatrix}\begin{pmatrix} -d' & c'\sqrt{n} \\ b'\sqrt{n} & -a' \end{pmatrix}\begin{pmatrix} 0 & -1 \\ 1 & 0 \end{pmatrix} = \begin{pmatrix} a' & b'\sqrt{n} \\ c'\sqrt{n} & d' \end{pmatrix} \in \Gamma_2,$$

with $|a'| < |b'|\sqrt{n} < |a|$. If $a' = 0$, then as before $\begin{pmatrix} a' & b'\sqrt{n} \\ c'\sqrt{n} & d' \end{pmatrix} \in \Gamma_1$. If $a' \neq 0$, then we repeat the entire process to obtain $\begin{pmatrix} a'' & b''\sqrt{n} \\ c''\sqrt{n} & d'' \end{pmatrix} \in \Gamma_2$, such that $|a''| < |a'| < |a|$ or $b'' = 0$. Since $a,a',a'' \in Z$ we eventually get an element in Γ_2 of one of the two forms $\begin{pmatrix} 0 & \beta\sqrt{n} \\ \gamma\sqrt{n} & \delta \end{pmatrix}$ or $\begin{pmatrix} \alpha & 0 \\ \gamma\sqrt{n} & \delta \end{pmatrix}$. Both of these are in Γ_1. Since all multiplications are by elements of Γ_1, we conclude that $\begin{pmatrix} a & b\sqrt{n} \\ c\sqrt{n} & d \end{pmatrix} \in \Gamma_1$, so that all elements in Γ_2 of type (a) are in Γ_1.

If $\begin{pmatrix} a\sqrt{n} & b \\ c & d\sqrt{n} \end{pmatrix} \in \Gamma_2$, then $\begin{pmatrix} a\sqrt{n} & b \\ c & d \end{pmatrix}\begin{pmatrix} 0 & -1 \\ 1 & 0 \end{pmatrix} = \begin{pmatrix} b & -a\sqrt{n} \\ d\sqrt{n} & -c \end{pmatrix} \in \Gamma_1$, so that $\begin{pmatrix} a\sqrt{n} & b \\ c & d\sqrt{n} \end{pmatrix} \in \Gamma_1$. Hence $\Gamma_2 \subset \Gamma_1$, and the proof is complete.

Corollary 3. $\Gamma(1)$ is generated by S and T.

Proof. This is the case $n = 1$ of Theorem 2.

Corollary 4. Γ_ϑ consists of all linear fractional transformations of the form $\tau' = (a\tau + b)/(c\tau + d)$; $a,b,c,d \in Z$, $ad - bc = 1$ with $\begin{pmatrix} a & b \\ c & d \end{pmatrix} \equiv \begin{pmatrix} 1 & 0 \\ 0 & 1 \end{pmatrix}$ or $\begin{pmatrix} 0 & -1 \\ 1 & 0 \end{pmatrix}$ (modulo 2). Here the congruence is elementwise.

Proof. This is the case $n = 4$ of Theorem 2.

We now obtain coset decompositions for $\Gamma(1)$ modulo $\Gamma^0(p)$, p a prime, and $\Gamma(1)$ modulo Γ_ϑ.

Theorem 5. As a coset decomposition for $\Gamma(1)$ modulo $\Gamma^0(p)$, p a prime, we may choose $S^j (0 \leqslant j \leqslant p - 1)$, T, in the sense that

$$\Gamma(1) = \bigcup_{j=0}^{p-1} \Gamma^0(p)S^j \cup \Gamma^0(p)T.$$

Proof. A simple check shows that these $p + 1$ cosets are distinct modulo $\Gamma^0(p)$. For $S^jT^{-1} = \begin{pmatrix} 1 & j \\ 0 & 1 \end{pmatrix}\begin{pmatrix} 0 & 1 \\ -1 & 0 \end{pmatrix} = \begin{pmatrix} -j & 1 \\ -1 & 0 \end{pmatrix} \notin \Gamma^0(p)$.

Let $\begin{pmatrix} a & b \\ c & d \end{pmatrix} \in \Gamma(1)$. There are two cases.

(1) Suppose $p|a$. Then $\begin{pmatrix} -b & a \\ -d & c \end{pmatrix} \in \Gamma^0(p)$ and $\begin{pmatrix} -b & a \\ -d & c \end{pmatrix}\begin{pmatrix} 0 & -1 \\ 1 & 0 \end{pmatrix} = \begin{pmatrix} a & b \\ c & d \end{pmatrix}$.

(2) Suppose $p \nmid a$; then $(a,p) = 1$. Hence there exist $j(0 \leqslant j \leqslant p-1)$ and $x \in Z$ such that $ja + xp = b$. Then

$$\begin{pmatrix} a & b - aj \\ c & d - cj \end{pmatrix} \in \Gamma^0(p) \qquad \text{and} \qquad \begin{pmatrix} a & b - aj \\ c & d - cj \end{pmatrix}\begin{pmatrix} 1 & j \\ 0 & 1 \end{pmatrix} = \begin{pmatrix} a & b \\ c & d \end{pmatrix}.$$

Lemma 6. If Γ_1 and Γ_2 are conjugate subgroups of $\Gamma(1)$ [i.e., if there exists $B \in \Gamma(1)$ with $B\Gamma_1 B^{-1} = \Gamma_2$], then $[\Gamma(1):\Gamma_1] = [\Gamma(1):\Gamma_2]$. In fact, if $A_1,\ldots,A_\mu$ is a coset decomposition of $\Gamma(1)$ modulo $\Gamma_1 : \Gamma(1) = \Gamma_1 A_1 \cup \cdots \cup \Gamma_1 A_\mu$, then $BA_1,\ldots,BA_\mu$ is a coset decomposition of $\Gamma(1)$ modulo $\Gamma_2 : \Gamma(1) = \Gamma_2(BA_1) \cup \cdots \cup \Gamma_2(BA_\mu)$.

Proof. $\Gamma(1) = \Gamma_1 A_1 \cup \cdots \cup \Gamma_1 A_\mu = B^{-1}\Gamma_2 BA_1 \cup \cdots \cup B^{-1}\Gamma_2 BA_\mu$. Multiply by B on both sides to obtain $\Gamma(1) = \Gamma_2(BA_1) \cup \cdots \cup \Gamma_2(BA_\mu)$. We must show that $BA_1,\ldots,BA_\mu$ are distinct modulo Γ_2. Suppose $(BA_i)\cdot(BA_j)^{-1} \in \Gamma_2$; then $BA_iA_j^{-1}B^{-1} \in \Gamma_2$ or $A_iA_j^{-1} \in \Gamma_1$ and $i = j$, since $A_1,\ldots,A_\mu$ are distinct modulo Γ_1.

Corollary 7. As a coset decomposition of $\Gamma(1)$ modulo $\Gamma_0(p)$, p a prime, we may choose the $p + 1$ elements $TS^j(0 \leqslant j \leqslant p-1)$, $I = \begin{pmatrix} 1 & 0 \\ 0 & 1 \end{pmatrix}$.

Proof. A straightforward calculation shows that $\Gamma_0(p) = T\Gamma^0(p)T^{-1}$. The result then follows from Theorem 5 and Lemma 6.

Lemma 8. $\Gamma_\vartheta = S^{-1}\Gamma^0(2)S$.

Proof. To prove that $\Gamma_\vartheta \subset S^{-1}\Gamma^0(2)S$, it suffices to show that $STS^{-1} \in \Gamma^0(2)$ and $SS^2S^{-1} \in \Gamma^0(2)$. Clearly $SS^2S^{-1} = S^2 = \begin{pmatrix} 1 & 2 \\ 0 & 1 \end{pmatrix} \in \Gamma^0(2)$.

Now, $STS^{-1} = \begin{pmatrix} 1 & 1 \\ 0 & 1 \end{pmatrix}\begin{pmatrix} 0 & -1 \\ 1 & 0 \end{pmatrix}\begin{pmatrix} 1 & -1 \\ 0 & 1 \end{pmatrix} = \begin{pmatrix} 1 & -2 \\ 1 & -1 \end{pmatrix} \in \Gamma^0(2)$.

We next show that $S^{-1}\Gamma^0(2)S \subset \Gamma_\vartheta$, to conclude the proof. Given $\begin{pmatrix} a & b \\ c & d \end{pmatrix} \in \Gamma^0(2)$, we have $b \equiv 0 \pmod 2$. Then $a \equiv d \equiv 1 \pmod 2$ and c is $\equiv 0$ or $1 \pmod 2$. Now

$$\begin{pmatrix} 1 & -1 \\ 0 & 1 \end{pmatrix}\begin{pmatrix} a & b \\ c & d \end{pmatrix}\begin{pmatrix} 1 & 1 \\ 0 & 1 \end{pmatrix} = \begin{pmatrix} a - c & a - c + b - d \\ c & c + d \end{pmatrix}.$$

If $c \equiv 0 \pmod 2$, then $a - c \equiv 1$, $a - c + b - d \equiv 0$, $c \equiv 0$, $c + d \equiv 1$

(mod 2). If $c \equiv 1 \pmod 2$, then $a - c \equiv 0$, $a - c + b - d \equiv 1$, $c \equiv 1$, $c + d \equiv 0 \pmod 2$. In either case, by Corollary 4, $\begin{pmatrix} 1 & -1 \\ 0 & 1 \end{pmatrix}\begin{pmatrix} a & b \\ c & d \end{pmatrix}\cdot\begin{pmatrix} 1 & 1 \\ 0 & 1 \end{pmatrix} \in \Gamma_{\vartheta}$. Hence $S^{-1}\Gamma^0(2)S \subset \Gamma_{\vartheta}$ and $S^{-1}\Gamma^0(2)S = \Gamma_{\vartheta}$.

Corollary 9. As a coset decomposition of $\Gamma(1)$ modulo Γ_{ϑ} we may choose the three elements $I, S^{-1}, S^{-1}T$. Thus Γ_{ϑ} is of index 3 in $\Gamma(1)$.

Proof. This follows directly from Lemma 8, Theorem 5, and Lemma 6.

Theorem 10. Let p be a prime. Then $\Gamma_0(p^2)$ is of index p in $\Gamma_0(p)$ and as a coset decomposition of $\Gamma_0(p)$ modulo $\Gamma_0(p^2)$ we may write

$$\Gamma_0(p) = \bigcup_{k=0}^{p-1} \Gamma_0(p^2)\begin{pmatrix} 1 & 0 \\ -pk & 1 \end{pmatrix}.$$

Proof. Clearly the elements $\begin{pmatrix} 1 & 0 \\ -pk & 1 \end{pmatrix}$, $0 \leqslant k \leqslant p - 1$, are in distinct cosets modulo $\Gamma_0(p^2)$. Hence the index is at least p. We must show that every element of $\Gamma_0(p)$ is in one of the sets $\Gamma_0(p^2)\begin{pmatrix} 1 & 0 \\ -pk & 1 \end{pmatrix}, 0 \leqslant k \leqslant p - 1$.

Let $\begin{pmatrix} a & b \\ c & d \end{pmatrix} \in \Gamma_0(p)$, so that $p|c$. Write $c = p(pt + u)$, $0 \leqslant u \leqslant p - 1$. Observe that $ad - bc = 1$ implies that $(p,d) = 1$. Now consider

$$\begin{pmatrix} a & b \\ c & d \end{pmatrix}\begin{pmatrix} 1 & 0 \\ pk & 1 \end{pmatrix} = \begin{pmatrix} a & b \\ p(pt + u) & d \end{pmatrix}\begin{pmatrix} 1 & 0 \\ pk & 1 \end{pmatrix} = \begin{pmatrix} * & * \\ p(pt + u + dk) & * \end{pmatrix}.$$

But $(p,d) = 1$ implies that there exists k such that $0 \leqslant k \leqslant p - 1$ with $dk \equiv -u \pmod p$; that is, $p|(dk + u)$. With this choice of k,

$$\begin{pmatrix} a & b \\ c & d \end{pmatrix}\begin{pmatrix} 1 & 0 \\ pk & 1 \end{pmatrix} \in \Gamma_0(p^2), \text{ or } \begin{pmatrix} a & b \\ c & d \end{pmatrix} \in \Gamma_0(p^2)\begin{pmatrix} 1 & 0 \\ -pk & 1 \end{pmatrix}.$$

The proof is complete.

4. FUNDAMENTAL REGIONS OF SUBGROUPS

Lemma 11. Let $V \in \Gamma(1)$. If there is $\tau_0 \in \mathscr{R}(\Gamma(1))$ such that $V\tau_0 = \tau_0$, then $V = I$.

Proof. Since $V\tau_0 = \tau_0$ and V is continuous at $\tau_0 \in \mathscr{H}$, there are neighborhoods N_1, N_2 of τ_0 such that $N_1 \subset N_2 \subset \mathscr{R}(\Gamma(1))$ and $V(N_1) \subset N_2 \subset \mathscr{R}(\Gamma(1))$. Since $\mathscr{R}(\Gamma(1))$ is a F.R. for $\Gamma(1)$ it follows that $V\tau = \tau$ for all $\tau \in N_1$. Since V is regular on $\mathscr{H}$, $V = I$.

Theorem 12. Let Γ be a subgroup of $\Gamma(1)$, with cosets $A_1, A_2, \ldots, A_\mu$ in the sense that $\Gamma(1) = \bigcup_{i=1}^{\mu} \Gamma A_i$. Then as a F.R. for Γ we may choose $R = \bigcup_{i=1}^{\mu} A_i\{\mathscr{R}(\Gamma(1))\}$.

Definition. We will call a F.R. $\mathscr{R}$ of this form a *standard fundamental region* (S.F.R.) for Γ.

Proof. We first prove that no two distinct points of $\mathscr{R}$ are equivalent with respect to Γ. Toward this end suppose that τ_1 and τ_2 are in $\mathscr{R}$ and equivalent with respect to Γ. Then $\tau_1 = A_i x$, $\tau_2 = A_j y$, and $\tau_2 = M\tau_1$, where x and y are in $R(\Gamma(1))$, $1 \leqslant i, j \leqslant \mu$, and $M \in \Gamma$. If follows that $A_j y = MA_i x$, or $y = A_j^{-1}MA_i x$. But $A_j^{-1}MA_i \in \Gamma(1)$; hence $x = y$ since $\mathscr{R}(\Gamma(1))$ is a F.R. for $\Gamma(1)$. By Lemma 11, $A_j^{-1}MA_i = I$, or $A_jA_i^{-1} = M \in \Gamma$. This implies that $i = j$ since the A's are cosets and therefore distinct modulo Γ. Hence $\tau_1 = A_i x = A_j y = \tau_2$, so that no two distinct points of $\mathscr{R}$ are equivalent with respect to Γ.

We now show that any point of $\mathscr{H}$ is equivalent with respect to Γ to a point in $\bar{\mathscr{R}}$. Given any $\tau_0 \in \mathscr{H}$, there exists $V \in \Gamma(1)$ such that $\tau_0 = Vx$, where $x \in \overline{\mathscr{R}(\Gamma(1))}$. On the other hand, there exist $M \in \Gamma$ and $i (1 \leqslant i \leqslant \mu)$ such that $V = MA_i$. Hence $\tau_0 = MA_i x$, or $M^{-1}\tau_0 = A_i x \in A_i(\overline{\mathscr{R}(\Gamma(1))})$. Since A_i is actually a homeomorphism of $\mathscr{H}$ onto itself, it is easy to see that $A_i(\overline{\mathscr{R}(\Gamma(1))}) = \overline{A_i(\mathscr{R}(\Gamma(1)))} \subset \bar{\mathscr{R}}$. Hence $M^{-1}\tau_0 \in \bar{\mathscr{R}}$, and we are done.

Corollary 13. As a F.R. for $\Gamma_0(p)$, p prime, we may choose

$$\mathscr{R}(\Gamma(1)) \bigcup_{j=0}^{p-1} TS^j\{\mathscr{R}(\Gamma(1))\}.$$

Proof. By Corollary 7 and Theorem 12.

Corollary 14. As a F.R. for Γ_ϑ we may choose

$$\mathscr{R}(\Gamma(1)) \cup S^{-1}\{\mathscr{R}(\Gamma(1))\} \cup S^{-1}T\{\mathscr{R}(\Gamma(1))\}.$$

Proof. This follows from Corollary 9 and Theorem 12.

Combining Theorem 12 with Lemma 6, we obtain the more general

Corollary 15. If Γ_1, Γ_2 are conjugate subgroups of finite index in $\Gamma(1)$, with $B\Gamma_1B^{-1} = \Gamma_2$, $B \in \Gamma(1)$ and $\mathscr{R}_1$ is a S.F.R. for Γ_1, then $B(\mathscr{R}_1)$ is a S.F.R. for Γ_2.

We close this chapter with one more result concerning S.F.R.'s for subgroups of $\Gamma(1)$.

Theorem 16. Suppose $\Gamma_2 \subset \Gamma_1 \subset \Gamma(1)$, where Γ_1 and Γ_2 are subgroups of finite index in $\Gamma(1)$. If $\mathscr{R}_1$ is a S.F.R. for Γ_1 and if we have the coset decomposition $\Gamma_1 = \bigcup_{i=1}^{\mu} \Gamma_2 A_i$, then $\mathscr{R}_2 = \bigcup_{i=1}^{\mu} A_i(\mathscr{R}_1)$ is a S.F.R. for Γ_2.

Proof. The proof of Theorem 12 shows that $\mathscr{R}_2$ is a F.R. for Γ_2. One needs only to replace $\Gamma(1)$ by Γ_1 and Γ by Γ_2 in that proof. We must yet show that $\mathscr{R}_2$ is a S.F.R. For this purpose write $\mathscr{R}_1 = \bigcup_{j=1}^{\nu} B_j\{\mathscr{R}(\Gamma(1))\}$, where $\Gamma(1) = \bigcup_{j=1}^{\nu} \Gamma_1 B_j$ is a coset decomposition of $\Gamma(1)$ modulo Γ_1. This expression for $\mathscr{R}_1$ exists since $\mathscr{R}_1$ is assumed to be a S.F.R. for Γ_1. Then, of course,

$$\mathscr{R}_2 = \bigcup_{i=1}^{\mu} A_i\left(\bigcup_{j=1}^{\nu} B_j\{\mathscr{R}(\Gamma(1))\}\right) = \bigcup_{i=1}^{\mu}\bigcup_{j=1}^{\nu} A_i B_j\{\mathscr{R}(\Gamma(1))\}.$$

If we can show that $\Gamma(1) = \bigcup_{i=1}^{\mu}\bigcup_{j=1}^{\nu} \Gamma_2 A_i B_j$ is a genuine coset decomposition of $\Gamma(1)$ modulo Γ_2, then we shall be done.

Since $\Gamma(1) = \bigcup_{j=1}^{\nu} \Gamma_1 B_j$ and $\Gamma_1 = \bigcup_{i=1}^{\mu} \Gamma_2 A_i$, we have

$$\Gamma(1) = \bigcup_{j=1}^{\nu}\bigcup_{i=1}^{\mu} \Gamma_2 A_i B_j = \bigcup_{i=1}^{\mu}\bigcup_{j=1}^{\nu} \Gamma_2 A_i B_j,$$

where thus far this is only a set equality. To finish the proof we show that the elements $\{A_i B_j | 1 \leqslant i \leqslant \mu, 1 \leqslant j \leqslant \nu\}$ are distinct modulo Γ_2. Suppose $A_i B_j$ and $A_k B_l$ are in the same coset modulo Γ_2. Then $(A_i B_j)(A_k B_l)^{-1} \in \Gamma_2$ and, since $\Gamma_2 \subset \Gamma_1$, $B_j B_l^{-1} \in A_i^{-1}\Gamma_2 A_k \subset \Gamma_1$. Since the B's are distinct modulo Γ_1 by assumption, we have $j = l$ and $A_i A_k^{-1} \in \Gamma_2$. Since the A's are distinct modulo Γ_2, we also have $i = k$. This completes the proof.

Chapter 2

MODULAR FUNCTIONS AND FORMS

1. MULTIPLIER SYSTEMS

This chapter introduces the important concepts of "modular function" and "modular form" and presents those results (Section 5) that are the basis for our applications of the theory of modular functions and forms to number theory. The introduction of these concepts, however, requires a certain amount of preliminary discussion and we, in fact, do not give the complete definitions until Section 4. Throughout this chapter Γ denotes a subgroup of finite index in $\Gamma(1)$ and r a real number.

Roughly speaking, a modular form of degree† $-r$, with respect to Γ, is a function $F(\tau)$, defined and meromorphic in $\mathscr{H}$, which satisfies a growth condition at certain points of the real axis and the functional equations

$$(1) \qquad F(M\tau) = \upsilon(M)(c\tau + d)^r F(\tau), \qquad \tau \in \mathscr{H},$$

for all $M = \begin{pmatrix} * & * \\ c & d \end{pmatrix} \in \Gamma$. Here $\upsilon(M)$ is a complex number, independent of τ, such that $|\upsilon(M)| = 1$ for all $M \in \Gamma$. We fix the branch of $(c\tau + d)^r$ by adopting the following convention to which, unless otherwise noted, we adhere throughout the book. If z is any complex number, put

$$(2) \qquad z^r = |z|^r e^{ir \arg(z)}, \qquad -\pi \leqslant \arg(z) < \pi.$$

† Until recently most writers on modular functions and forms (including the present writer) used the term "dimension" rather than "degree."

Note that, according to (2), $\arg(-1) = -\pi$. It is easy to prove that if there exists $F(\tau) \not\equiv 0$ satisfying (1), then for any $M_1, M_2 \in \Gamma$,

$$(3) \qquad v(M_1M_2)(c_3\tau + d_3)^r = v(M_1)v(M_2)(c_1M_2\tau + d_1)^r(c_2\tau + d_2)^r,$$

where $M_1 = \begin{pmatrix} * & * \\ c_1 & d_1 \end{pmatrix}$, $M_2 = \begin{pmatrix} * & * \\ c_2 & d_2 \end{pmatrix}$, and $M_3 = M_1M_2 = \begin{pmatrix} * & * \\ c_3 & d_3 \end{pmatrix}$.
It is of interest to observe that if r is an *integer*, then (3) reduces to $v(M_1M_2) = v(M_1)v(M_2)$; that is, v is a character on Γ in this case.

For the moment we drop the discussion of functions $F(\tau)$ satisfying (1) and concentrate instead on functions $v(M)$, defined on Γ, that satisfy (3). We give the

Definition. We say that v is a *multiplier system for the group* Γ *and the degree* $-r$ provided $v(M), M \in \Gamma$, is a complex-valued function of absolute value 1, satisfying equation (3).

Remarks. 1. In general v is a function on the *matrix group* corresponding to Γ rather than on the linear fractional group Γ itself. This can be seen as follows. Apply (3) with $M_1 = M_2 = I$ to get $v(I)^2 = v(I)$; that is, since $v(I) \neq 0$, $v(I) = 1$. Apply (3) again with $M_1 = M_2 = -I$ to get $v(I) = v(-I)^2(-1)^{2r}$, $\arg(-1) = -\pi$. Hence $v(-I)^2 = e^{2\pi ir}$, and if r is not an integer $v(-I) \neq v(I)$. Thus the function v distinguishes between the matrices M and $-M$, although these correspond to the same linear fractional transformation.

2. In order that there exist $F(\tau) \not\equiv 0$ satisfying (1) we must have $v(I) = 1$ and $v(-I) = e^{\pi ir}$. For $F(\tau) = F(I\tau) = v(I)F(\tau)$; hence $v(I) = 1$. Also, $F(\tau) = F(-I\tau) = v(-I)(-1)^rF(\tau) = v(-I)e^{-\pi ir}F(\tau)$; hence $v(-I) = e^{\pi ir}$.

From these two equations and (3) one can easily deduce that $v(-M)\cdot(-c\tau - d)^r = v(M)(c\tau + d)^r$, for all $M \in \Gamma$. This latter equality can also be deduced directly from the existence of a nontrivial F satisfying (1).

2. PARABOLIC POINTS

In Section 1 we referred to a "growth condition at certain points of the real axis." In order to clarify this we must introduce and discuss the notion of a "parabolic point."

Definition. Let $\mathcal{R}$ be a F.R. of Γ. A *parabolic point* (or *parabolic vertex* or *parabolic cusp*) of Γ in $\mathcal{R}$ is any real point q, or $q = \infty$, such that $q \in \bar{\mathcal{R}}$, the closure of $\mathcal{R}$ in the topology of the Riemann sphere.

Remarks. 1. Let $\mathcal{R}$ be the S.F.R. of Γ constructed in Theorem 12 of Chapter 1. We shall show that all the parabolic points of $\mathcal{R}$ are *rational points*, where now and henceforth we adopt the useful convention that $\infty = 1/0$ is a *rational point*. By Theorem 1 of Chapter 1, the only real point in the closure of $\mathscr{R}(\Gamma(1))$ is ∞. Thus the only real points in the closure of $\mathscr{R}$ are the points $A_1(\infty), A_2(\infty), \ldots, A_\mu(\infty)$. Since $A_1, A_2, \ldots, A_\mu \in \Gamma(1)$, it follows that the points $q_i = A_i(\infty)$, $1 \leqslant i \leqslant \mu$, are rational points.

2. The preceding remark shows that the parabolic points of the S.F.R. $\mathscr{R}$ are precisely the points $A_1(\infty), A_2(\infty), \ldots, A_\mu(\infty)$. However, we should also take note of the fact that the points $q_i = A_i(\infty)$, $1 \leqslant i \leqslant \mu$, need not be distinct. It might (and often does—see Examples 1, 2, and 3, below) occur that $q_i = q_j$, with $i \neq j$. In this case $A_j^{-1}A_i(\infty) = \infty$, and a simple calculation shows that $A_j^{-1}A_i = S^n$, for some integer n. We conclude that $A_i = A_jS^n$ and, in particular, $A_i\{\mathscr{R}(\Gamma(1))\} = A_jS^n\{\mathscr{R}(\Gamma(1))\}$. This leads to the following simple, but important, lemma.

Lemma 1. Let $\mathscr{R}$ be a S.F.R. for Γ of the form $R = \bigcup_{i=1}^{\mu} A_i\{\mathscr{R}(\Gamma(1))\}$. Let $q_i = A_i(\infty)$, $1 \leqslant i \leqslant \mu$, be the parabolic points of Γ in $\mathscr{R}$. Then if $\tau \to q_i$ from within $\mathscr{R}$, it follows that $A_i^{-1}(\tau) \to \infty$ from within a vertical strip of the form $\operatorname{Im} z > 0$, $a_i < \operatorname{Re} z < b_i$, with a_i, b_i real numbers.

Proof. Since $A_i^{-1}(q_i) = \infty$ and A_i is a linear fractional transformation, it follows that $A_i^{-1}(\tau) \to \infty$ as $\tau \to q_i$. Suppose $q_i < \infty$. Then there exists a nonempty open disc D centered at q_i such that

$$D \cap \mathscr{R} = D \cap \left(\bigcup_{\nu \in \sigma(i)} A_\nu\{\mathscr{R}(\Gamma(1))\} \right),$$

where $\sigma(i) = \{\nu | 1 \leqslant \nu \leqslant \mu \text{ and } A_\nu(\infty) = A_i(\infty), \text{ that is, } q_\nu = q_i\}$. For each $\nu \in \sigma(i)$ there exists an integer $n(\nu)$ such that $A_\nu = A_iS^{n(\nu)}$. Thus if $\tau \in D \cap \mathscr{R}$, then $A_i^{-1}(\tau)$ remains within $A_i^{-1}(D) \cap (\bigcup_{\nu \in \sigma(i)} S^{n(\nu)}\{\mathscr{R}(\Gamma(1))\})$, which, in turn, is contained in a vertical strip of the form $\operatorname{Im} z > 0$, $a_i < \operatorname{Re} z < b_i$, with a_i and b_i real numbers. Thus as $\tau \to q_i$ from within $\mathscr{R}$, $A_i^{-1}(\tau) \to \infty$ from within this vertical strip. If $q_i = \infty$, the proof is the same, except that the disc D must be replaced by a half-plane of the form $\operatorname{Im} z > y_0$, with $y_0 > 0$.

Examples. From the coset decompositions given in Chapter 1 we can now determine the parabolic points of a S.F.R. for each of the groups Γ_ϑ, $\Gamma_0(p)$, and $\Gamma_0(p^2)$, with p a prime number.

1. By Corollary 9 of Chapter 1 we may choose $I, S^{-1}, S^{-1}T$ as a set of coset representatives for $\Gamma(1)$ modulo Γ_ϑ. $I(\infty) = S^{-1}(\infty) = \infty$ and $S^{-1}T(\infty) = S^{-1}(0) = -1$. Thus ∞ and -1 are the parabolic points of Γ_ϑ corresponding to the S.F.R. in question.

2. By Corollary 7 of Chapter 1 the parabolic points of a S.F.R. of $\Gamma_0(p)$ are ∞ and $TS^j(\infty)$, $0 \leqslant j \leqslant p-1$. But $TS^j(\infty) = T(\infty) = 0$. Thus ∞ and 0 are the parabolic points.

3. By Theorems 10 and 16 of Chapter 1 and Example 2 (above) the parabolic points of a S.F.R. for $\Gamma_0(p^2)$ are $W^{-pk}(\infty)$ and $W^{-pk}(0)$,

$0 \leqslant k \leqslant p-1$, where $W = \begin{pmatrix} 1 & 0 \\ 1 & 1 \end{pmatrix}$. Now $W^{-pk}(0) = 0$ for $0 \leqslant k \leqslant p-1$, and $W^{-pk}(\infty) = -1/pk$, $1 \leqslant k \leqslant p-1$. Of course $W^0(\infty) = I(\infty) = \infty$. Thus the parabolic points of a S.F.R. for $\Gamma_0(p^2)$ are ∞, 0, and $-1/pk$, $1 \leqslant k \leqslant p-1$.

Lemma 2. Let q be any rational point and let $\Gamma_q = \{M \in \Gamma | M(q) = q\}$. Then Γ_q is a nontrivial cyclic subgroup of Γ. Furthermore, each $M = \begin{pmatrix} \alpha & \beta \\ \gamma & \delta \end{pmatrix} \in \Gamma_q$ can be so normalized that $\alpha + \delta = 2$.

Proof. We claim first that there exists a positive integer n such that $S^n \in \Gamma$. For otherwise $S, S^2, S^3,\ldots$ represent infinitely many distinct cosets of $\Gamma(1)$ modulo Γ. We have assumed that Γ is of finite index in $\Gamma(1)$.

Clearly Γ_q is a subgroup of Γ. We must show that Γ_q is nontrivial and cyclic. Suppose that $q = \infty$. By the previous paragraph we choose the smallest positive integer n such that $S^n \in \Gamma$. Of course $S^n \in \Gamma_\infty$, so that Γ_∞ is nontrivial. We claim also that S^n generates all of Γ_∞. Suppose $M \in \Gamma_\infty$; then $M(\infty) = \infty$, and we conclude that $M = S^\lambda$ for some integer λ. If $\lambda = 0$, then $M = I$ and there is nothing to prove. If $\lambda < 0$, replace M by $M^{-1} = S^{-\lambda}$. Thus we may assume that $\lambda > 0$. By the definition of n, $\lambda \geqslant n$. Using the division algorithm we write $\lambda = nt + r, 0 \leqslant r < n$. Since $S^\lambda, S^n \in \Gamma_\infty$, it follows that $S^r \in \Gamma_\infty$, so that by the definition of n, $r = 0$. Thus $\lambda = nt$, and we conclude that $M = S^\lambda = (S^n)^t$. In any case, M is a power of S^n and it follows that Γ_∞ is a cyclic subgroup of Γ.

Suppose that $q < \infty$ and write $q = a/b$, with a and b integers such that $(a,b) = 1$. Then there exists $V = \begin{pmatrix} x & y \\ b & -a \end{pmatrix} \in \Gamma(1)$, for $(a,b) = 1$ implies the existence of integers x and y such that $-xa - by = 1$. Clearly $V(a/b) = \infty$, and $V\Gamma_q V^{-1}$ is the subgroup of $V\Gamma V^{-1}$ leaving ∞ fixed. By the earlier part of the proof, $V\Gamma_q V^{-1}$ is a nontrivial cyclic subgroup of $V\Gamma V^{-1}$. Thus Γ_q is a nontrivial cyclic subgroup of Γ.

If $q = \infty$, then every element M of Γ_∞ has the form $M = (S^n)^t = \begin{pmatrix} 1 & nt \\ 0 & 1 \end{pmatrix}$. If $q < \infty$, write $q = a/b$, as above. Then a short calculation shows that every $M \in \Gamma_q$ has the form $M = V^{-1} S^{n(q)t} V = \begin{pmatrix} 1 - abn(q)t & a^2 n(q)t \\ -b^2 n(q)t & 1 + abn(q)t \end{pmatrix}$, where $t \in Z$ and $n(q)$ is the minimal positive integer such that $S^{n(q)} \in V\Gamma V^{-1}$, so that $S^{n(q)}$ is a generator of the cyclic group $V\Gamma_q V^{-1}$. In either case, if we write $M = \begin{pmatrix} \alpha & \beta \\ \gamma & \delta \end{pmatrix}$, then $\alpha + \delta = 2$. For convenience we also write $n = n(\infty)$.

Remarks. 1. A matrix or linear fractional transformation $M = \begin{pmatrix} \alpha & \beta \\ \gamma & \delta \end{pmatrix} \neq I$ such that $\alpha + \beta = \pm 2$ is called *parabolic*. Thus the results of this section imply that every parabolic point of S.F.R. for Γ is left fixed by a nontrivial cyclic subgroup of Γ, every element of which is parabolic. Since $S^n \neq I$ for every nonzero integer n, it follows that each Γ_q is, in fact, an *infinite* cyclic subgroup of Γ.

2. It is of interest, and also of some importance, to verify that the positive integer $n(q)$ occurring at the end of the proof of Lemma 2 depends only on q and *not* on the choice of V. To verify this, suppose $V_1(q) = \infty = V_2(q)$. Then $V_2 V_1^{-1}(\infty) = \infty$, so that $V_2 = S^\lambda V_1$ and $V_2 \Gamma V_2^{-1} = S^\lambda V_1 \Gamma V_1^{-1} S^{-\lambda}$, for some integer λ. Thus the subgroup of $V_2 \Gamma V_2^{-1}$ leaving ∞ fixed is precisely the same group as the subgroup of $V_1 \Gamma V_1^{-1}$ leaving ∞ fixed. It follows that $n(q)$ is independent of the choice of $V(V(q) = \infty)$.

3. Suppose $\mathscr{R}$ is a S.F.R. for Γ with parabolic points $q_j = A_j(\infty), 1 \leqslant j \leqslant \mu$. The q_j are rational points, and for V we may of course choose A_j^{-1}. In view of Remark 2 (above) this choice has no effect on the integer $n(q_j)$. Thus we have the following

Definition. If $q_j = A_j(\infty)$, $1 \leqslant j \leqslant \mu$, are the parabolic points in a S.F.R. $\mathscr{R}$ of Γ, we put $\lambda_j = n(q_j)$, the smallest positive integer such that $S^{\lambda_j} \in A_j^{-1} \Gamma A_j$. We call λ_j the *width of the parabolic point* q_j. In view of the remarks preceding this definition, if $q_i = A_i(\infty) = A_j(\infty) = q_j$, then $\lambda_i = \lambda_j$. This shows that the "width of a parabolic point" is a well-defined concept, depending only on the parabolic point itself, and not on any particular representation of the point in the form $V(\infty), V \in \Gamma(1)$. Another kind of invariance of "width" is contained in

Proposition 3. Suppose

$$\mathscr{R} = \bigcup_{j=1}^{\mu} A_j\{\mathscr{R}(\Gamma(1))\} \text{ and } \mathscr{R}' = \bigcup_{j=1}^{\mu} A_j'\{\mathscr{R}(\Gamma(1))\}$$

are two S.F.R.'s of Γ, with the numbering so arranged that $A_j' \in \Gamma A_j$, for $1 \leqslant j \leqslant \mu$. Let the parabolic points of $\mathscr{R}$ be q_j of width λ_j, and let those of $\mathscr{R}'$ be q_j' of width λ_j', $1 \leqslant j \leqslant \mu$. Then $\lambda_j' = \lambda_j$ for $1 \leqslant j \leqslant \mu$.

Proof. We may write $A_j' = W A_j$, where $W \in \Gamma$, depending on j. Thus $A_j'^{-1} \Gamma A_j' = A_j^{-1} W^{-1} \Gamma W A_j = A_j^{-1} \Gamma A_j$, and we conclude that the subgroup of $A_j'^{-1} \Gamma A_j'$ leaving ∞ fixed is precisely the same group as the subgroup of $A_j^{-1} \Gamma A_j$ leaving ∞ fixed. Thus $\lambda_j' = \lambda_j$, and the proof is complete.

Proposition 3 implies that the width of a parabolic point of q of Γ is determined by the equivalence class of q modulo Γ. That is, if q is a parabolic point in a S.F.R. of Γ and $W \in \Gamma$, then $q' = Wq$ is also a parabolic point in some S.F.R. of Γ, and the width of q' equals the width of q.

3. FOURIER EXPANSIONS

We fix a S.F.R. $\mathscr{R} = \bigcup_{i=1}^{\mu} A_i\{\mathscr{R}(\Gamma(1))\}$ of Γ, with parabolic points $q_j = A_j(\infty)$, $1 \leqslant j \leqslant \mu$. If $F(\tau)$ is *meromorphic* in $\mathscr{H}$ and satisfies (1), for all $M \in \Gamma$, we shall show that $F(\tau)$ has infinite series expansions of a special type "at" the points $q_1,\dots,q_\mu$. These expansions, which in fact are essentially Laurent expansions, will be referred to as "Fourier expansions" since they have an exponential form. To describe the expansions precisely we need the following

Definition. Suppose v is a fixed multiplier system for the group Γ and the degree $-r$. For $1 \leqslant j \leqslant \mu$ we define the real number κ_j by means of

$$(4) \qquad v(A_j S^{\lambda_j} A_j^{-1}) = e^{2\pi i \kappa_j}, \qquad 0 \leqslant \kappa_j < 1.$$

Note that $A_j S^{\lambda_j} A_j^{-1} \in \Gamma$, so that (4) is meaningful. Also, in view of the remarks following the proof of Lemma 2, if $q_i = q_j$, with $i \neq j$, it follows that $A_i S^{\lambda_i} A_i^{-1} = A_j S^{\lambda_j} A_j^{-1}$. Thus we have $\kappa_i = \kappa_j$, so that (4) defines a number depending only upon the parabolic point q_j and not upon its representation.

We are now in a position to state

Theorem 4. Suppose $F(\tau)$ is meromorphic in $\mathscr{H}$, satisfies (1) for all $M \in \Gamma$, and, in addition, has only finitely many poles in $\bar{\mathscr{R}} \cap \mathscr{H}$. Then for each $j = 1,\dots,\mu$, there exists a nonnegative real number y_j such that $F(\tau)$ has the following expansion, valid for $\operatorname{Im}(A_j^{-1}\tau) > y_j$:

$$(5) \qquad F(\tau) = \sigma_j(\tau) \sum_{n=-\infty}^{\infty} a_n(j) e^{2\pi i (n+\kappa_j)(A_j^{-1}\tau)/\lambda_j},$$

where λ_j is the width of q_j, κ_j is defined by (4), the $a_n(j)$ are complex numbers, and

$$\sigma_j(\tau) = \begin{cases} 1 & \text{if } q_j = \infty, \\ (\tau - q_j)^{-r} & \text{if } q_j < \infty. \end{cases}$$

The expansion (5) is called the "Fourier expansion of $F(\tau)$ at q_j." If $F(\tau)$ is actually regular in $\mathscr{H}$, then the expansions (5) are valid in all of $\mathscr{H}$; that is, we may choose $y_j = 0$ for $1 \leqslant j \leqslant \mu$.

The following lemma will be useful in the proof of Theorem 4 and also later in this chapter.

Lemma 5. Suppose q is a rational point $\neq \infty$ and $M = \begin{pmatrix} * & * \\ \gamma & \delta \end{pmatrix} \in \Gamma(1)$ is such that $M(q) = q$. Then for any real number r, $(M\tau - q)^r = (\gamma\tau + \delta)^{-r}(\tau - q)^r$.

Proof. Since q is a finite rational point, write $q = a/b$, with a and b relatively prime integers, $b \neq 0$. As in the proof of Lemma 2, M has the form

$$M = \begin{pmatrix} 1 - ab\lambda & a^2\lambda \\ -b^2\lambda & 1 + ab\lambda \end{pmatrix},$$

with λ an integer. The desired equation then takes the form

$$\left(\frac{(1 - ab\lambda)\tau + a^2\lambda}{-b^2\lambda\tau + (1 + ab\lambda)} - \frac{a}{b}\right)^r = (-b^2\lambda\tau + 1 + ab\lambda)^{-r}(\tau - a/b)^r,$$

or, after a simplification,

$$\left(\frac{\tau - a/b}{-b^2\lambda\tau + (1 + ab\lambda)}\right)^r = (-b^2\lambda\tau + 1 + ab\lambda)^{-r}(\tau - a/b)^r. \tag{6}$$

Clearly both sides in (6) have the same absolute value. Thus it suffices to show that both sides have the same argument. Now

$$\arg\left(\frac{\tau - a/b}{-b^2\lambda\tau + 1 + ab\lambda}\right) = \arg(\tau - a/b) - \arg(-b^2\lambda\tau + 1 + ab\lambda) + 2n\pi,$$

for some integer n. We shall show that $n = 0$. We have

$$(\tau - a/b)/(-b^2\lambda\tau + 1 + ab\lambda) = M\tau - q \in \mathscr{H},$$

so that $0 < \arg\{(\tau - a/b)/(-b^2\lambda\tau + 1 + ab\lambda)\} < \pi$, in accordance with our convention (2). Also,

$$0 < \arg(\tau - a/b) < \pi \text{ and } -\pi < \arg(-b^2\lambda\tau + 1 + ab\lambda) < \pi.$$

We conclude that

$$2\pi|n| = \left|\arg\left(\frac{\tau - a/b}{-b^2\lambda\tau + 1 + ab\lambda}\right) - \arg(\tau - a/b) + \arg(-b^2\lambda\tau + 1 + ab\lambda)\right| < 2\pi,$$

and $n = 0$. It follows that

$$\arg\left(\frac{\tau - a/b}{-b^2\lambda\tau + 1 + ab\lambda}\right) = \arg(\tau - a/b) - \arg(-b^2\lambda\tau + 1 + ab\lambda).$$

Multiplying this equation by r, we find that the argument is the same on both sides of (6). Thus (6) follows and the proof is complete.

Proof of Theorem 4. To simplify the notation we replace q_j by q, λ_j by λ, κ_j by κ, and A_j by A everywhere in the proof. Observe first that because $F(\tau)$ satisfies (1) for all $M \in \Gamma$, the condition "$F(\tau)$ has only finitely many poles in $\bar{\mathscr{R}} \cap \mathscr{H}$" is independent of the choice of the S.F.R. $\mathscr{R}$. Thus since S^λ is the smallest positive power of S in Γ, we may choose $S^n, 0 \leqslant n \leqslant \lambda - 1$, among the coset representatives of $\Gamma(1)$ modulo Γ. Since $\overline{\mathscr{R}(\Gamma(1))}$ contains the strip $-\frac{1}{2} \leqslant \operatorname{Re} \tau \leqslant \frac{1}{2}$, $\operatorname{Im} \tau \geqslant 1$, it follows that $F(\tau)$ has only finitely many poles in the strip $-\frac{1}{2} \leqslant \operatorname{Re} \tau \leqslant \lambda - \frac{1}{2}$, $\operatorname{Im} \tau \geqslant 1$, of width λ.

Suppose $q = \infty$. Then $M = AS^\lambda A^{-1} = S^\lambda \in \Gamma$, and we apply (1) to get $F(\tau + \lambda) = F(S^\lambda \tau) = v(S^\lambda)F(\tau) = e^{2\pi i\kappa}F(\tau)$. Putting $g(\tau) = e^{-2\pi i\kappa\tau/\lambda}F(\tau)$, we have $g(\tau + \lambda) = g(\tau)$, for $\tau \in \mathscr{H}$. With $w = e^{2\pi i\tau/\lambda}$, the function $h(w) = h(e^{2\pi i\tau/\lambda}) = g(\tau)$ is defined and meromorphic for $0 < |w| < 1$, since $F(\tau)$ is meromorphic in $\mathscr{H}$. Since $F(\tau)$ has only finitely many poles in the strip $-\frac{1}{2} \leqslant \operatorname{Re} \tau \leqslant \lambda - \frac{1}{2}$, $\operatorname{Im} \tau \geqslant 1$, $h(w)$ has only finitely many poles for $0 < |w| < e^{-2\pi/\lambda}$, and, consequently, $h(w)$ is regular in a punctured disc of the form $0 < |w| < \rho$. Thus $h(w)$ has the Laurent expansion at 0, valid for $0 < |w| < \rho$, $h(w) = \Sigma_{n=-\infty}^{\infty} a_n w^n$. In terms of τ this expansion is $F(\tau) = \Sigma_{n=-\infty}^{\infty} a_n e^{2\pi i(n+\kappa)\tau/\lambda}$, for $\operatorname{Im} \tau > y_0$, where $y_0 = (\lambda/2\pi)\log(1/\rho) > 0$. Since $q = \infty$, A is of the form $A = S^t$, with t an integer. Thus the expansion can be rewritten

$$F(\tau) = \sum_{n=-\infty}^{\infty} e^{2\pi i(n+\kappa)t/\lambda} a_n e^{2\pi i(n+\kappa)(A^{-1}\tau)/\lambda}, \operatorname{Im} \tau > y_0,$$

which is precisely (5) in the case $q_j = \infty$. If $F(\tau)$ is regular in $\mathscr{H}$, then $h(w)$ is regular for $0 < |w| < 1$. Thus we can choose $\rho = 1$, so that $y_0 = 0$.

Suppose now that $q < \infty$. With λ the width of q, put $M = AS^\lambda A^{-1} \in \Gamma$. Here M is a parabolic linear fractional transformation such that $M(q) = q$. Put $\varphi(\tau) = (\tau - q)^r F(\tau)$. By (1) we have

$$\begin{aligned}\varphi(M\tau) &= (M\tau - q)^r F(M\tau) = (M\tau - q)^r v(M)(\gamma\tau + \delta)^r F(\tau)\\ &= (M\tau - q)^r(\gamma\tau + \delta)^r e^{2\pi i\kappa}F(\tau),\end{aligned}$$

where $M = \begin{pmatrix} * & * \\ \gamma & \delta \end{pmatrix}$. By Lemma 5,

$$\varphi(M\tau) = e^{2\pi i\kappa}(\tau - q)^r F(\tau) = e^{2\pi i\kappa}\varphi(\tau), \text{ or } \varphi(AS^\lambda A^{-1}\tau) = e^{2\pi i\kappa}\varphi(\tau).$$

Since this is valid for all $\tau \in \mathscr{H}$, we may replace τ by $A\tau$ to obtain $\varphi(AS^\lambda\tau) = e^{2\pi i\kappa}\varphi(A\tau)$, $\tau \in \mathscr{H}$. Putting $g(\tau) = e^{-2\pi i\kappa\tau/\lambda}\varphi(A\tau)$, we find that

$$g(\tau + \lambda) = g(S^\lambda\tau) = e^{-2\pi i\kappa(\tau+\lambda)/\lambda}\varphi(AS^\lambda\tau) = e^{-2\pi i\kappa\tau/\lambda}\varphi(A\tau) = g(\tau).$$

As before put $w = e^{2\pi i\tau/\lambda}$, and define the function $h(w)$ in $0 < |w| < 1$ by $h(w) = h(e^{2\pi i\tau/\lambda}) = g(\tau)$. In terms of the original function $F(\tau)$, we have $g(\tau) = e^{-2\pi i\kappa\tau/\lambda}(A\tau - q)^r F(A\tau)$. Since $F(\tau)$ is meromorphic in $\mathscr{H}$ the same is true of $g(\tau)$, so that $h(w)$ is meromorphic in the punctured disc $0 < |w| < 1$.

In addition, the fact that $F(\tau)$ has only finitely many poles in $\bar{\mathscr{R}} \cap \mathscr{H}$ implies that $g(\tau)$ has only finitely many poles in $A^{-1}(\bar{\mathscr{R}}) \cap \mathscr{H}$.

At this point it is important to observe that the transformation property (1) of $F(\tau)$ with respect to Γ is inherited by $g(\tau)$ in the form of a transformation property with respect to $A^{-1}\Gamma A$. Let $A^{-1}VA$, $V \in \Gamma$, be an arbitrary element of $A^{-1}\Gamma A$, with $V = \begin{pmatrix} * & * \\ c & d \end{pmatrix}$. Then, in fact, using (1) and the expression for $g(\tau)$ in terms of $F(\tau)$, we find that

$$(7) \qquad g(A^{-1}VA\tau) = \alpha(\tau)g(\tau), \qquad \tau \in \mathscr{H},$$

where

$$\alpha(\tau) = \frac{e^{2\pi i\kappa\tau/\lambda}e^{-2\pi i\kappa(A^{-1}VA\tau)/\lambda}v(V)(cA\tau + d)^r(VA\tau - q)^r}{(A\tau - q)^r},$$

a function which is regular and zero-free in $\mathscr{H}$. Now we know that $g(\tau)$ has only finitely many poles in $\overline{A^{-1}(\mathscr{R})} \cap \mathscr{H}$; on the other hand, by the results of Chapter 1, $A^{-1}(\mathscr{R})$ is a S.F.R. for $A^{-1}\Gamma A$. By (7) it follows that $g(\tau)$ has only finitely many poles in $\bar{\mathscr{R}}' \cap \mathscr{H}$, for *any* S.F.R. $\mathscr{R}'$ of $A^{-1}\Gamma A$. Since λ is the smallest positive integer such that $S^\lambda \in A^{-1}\Gamma A$, we may choose $S^n, 0 \leqslant n \leqslant \lambda - 1$, among the coset representatives of $\Gamma(1)$ modulo $A^{-1}\Gamma A$. From this it follows that $g(\tau)$ has only finitely many poles in the strip $-\frac{1}{2} \leqslant \operatorname{Re} \tau \leqslant \lambda - \frac{1}{2}, \operatorname{Im} \tau \geqslant 1$, so that $h(w)$ is regular in a punctured disc of the form $0 < |w| < \rho, \rho \leqslant 1$. We conclude that $h(w)$ has a Laurent expansion in powers of w, valid in this punctured disc. Expressed in terms of $F(\tau)$, this expansion has the form (5), valid for $\operatorname{Im}(A^{-1}\tau) > y_0 = (\lambda/2\pi) \log(1/\rho) > 0$. If $F(\tau)$ is regular in $\mathscr{H}$, then $h(w)$ is regular for $0 < |w| < 1$, and we may choose $\rho = 1$, that is, $y_0 = 0$. This concludes the proof.

Remarks. 1. The following converse of Theorem 4 is easy to establish, and we leave the proof as an "exercise for the reader." *If $F(\tau)$ is meromorphic in $\mathscr{H}$ and has an expansion of the form* (5) *for each parabolic point q_j of $\mathscr{R}$, valid in* $\operatorname{Im}(A_j^{-1}\tau) > y_j > 0$, *then $F(\tau)$ has at most finitely many poles in $\bar{\mathscr{R}} \cap \mathscr{H}$.* Note that in this converse $F(\tau)$ is not assumed to satisfy (1) for the group Γ. In virtue of the converse the existence of at most finitely many poles in $\mathscr{R} \cap \mathscr{H}$ and the validity of the expansions (5) are equivalent conditions for functions meromorphic in $\mathscr{H}$, which satisfy (1) for the group Γ.

2. If $q_k = q_j$ with $k \neq j$, then $F(\tau)$ has two expansions of the form (5) at the parabolic point q_j. By the uniqueness of the Laurent expansion in a given annulus, the coefficients must in fact be the same and this means that $a_n(j)$ and $a_n(k)$ are closely related. It is a simple matter to find the exact

relationship. Since $A_j(\infty) = q_j = q_k = A_k(\infty)$, we conclude that $A_j = A_k S^t$, for some integer t. Since $q_j = q_k$, we have $\sigma_j(\tau) = \sigma_k(\tau)$, $\lambda_j = \lambda_k$, and $\kappa_j = \kappa_k$. Using these facts together with the expansions (5) for indices j and k, we find that

$$a_n(j) = e^{2\pi i(n+\kappa_j)t/\lambda_j} a_n(k), \qquad \text{for all } n.$$

We turn to another question of invariance. Suppose $\mathscr{R}$ and $\mathscr{R}'$ are two S.F.R.'s for Γ, arranged as in Proposition 3. By Proposition 3 we already know that the width of the parabolic point q'_j in $\mathscr{R}'$ is the same as the width of the parabolic point q_j in $\mathscr{R}$. We of course define κ'_j as we did κ_j, that is, for $1 \leqslant j \leqslant \mu$, $v(A'_j S^{\lambda_j} A'^{-1}_j) = e^{2\pi i \kappa'_j}$, $0 \leqslant \kappa'_j < 1$.

Suppose $F(\tau) \not\equiv 0$ is meromorphic in $\mathscr{H}$ and satisfies (1) for all $M \in \Gamma$. Suppose also $F(\tau)$ has at most finitely many poles in $\bar{\mathscr{R}} \cap \mathscr{H}$.. As we observed at the beginning of the proof of Theorem 4, it follows that $F(\tau)$ has at most finitely many poles in $\bar{\mathscr{R}}' \cap \mathscr{H}$. Thus by Theorem 4, $F(\tau)$ has the expansion (5) at q_j and also the following expansion at q'_j, valid for $\mathrm{Im}(A'^{-1}_j \tau) > y'_j > 0$:

(8) $$F(\tau) = \sigma'_j(\tau) \sum_{n=-\infty}^{\infty} a'_n(j) e^{2\pi i(n+\kappa'_j)(A'^{-1}_j \tau)/\lambda_j}$$

where

$$\sigma'_j(\tau) = \begin{cases} 1 & \text{if } q'_j = \infty, \\ (\tau - q'_j)^{-r} & \text{if } q'_j < \infty. \end{cases}$$

In (8) we have already used the fact that $\lambda'_j = \lambda_j$. The question now arises: How are the expansions (5) and (8) related? The answer is contained in

Theorem 6. For $1 \leqslant j \leqslant \mu$, we have $\kappa'_j = \kappa_j$ and $a'_n(j) = \beta(j) a_n(j)$, for all n, where $\beta(j)$ is a nonzero constant independent of n.

Proof. We fix j, $1 \leqslant j \leqslant \mu$, and simplify the notation by writing $A, A', \lambda, q, q', \kappa, \kappa', \sigma(\tau)$, and $\sigma'(\tau)$ for $A_j, A'_j, \lambda_j, q_j, q'_j, \kappa_j, \kappa'_j, \sigma_j(\tau)$, and $\sigma'_j(\tau)$, respectively. There exists $M \in \Gamma$ such that $A' = MA$, so that $q' = A'(\infty) = MA(\infty) = Mq$. Write $M = \begin{pmatrix} a & b \\ c & d \end{pmatrix}$. Also put $V = AS^\lambda A^{-1}$; note that $V \in \Gamma$ and $Vq = q$.

It is convenient to begin by proving first a result similar to Lemma 5. This is

(9) $$(c\tau + d)^{-r} \sigma'(M\tau) = K\sigma(\tau),$$

where $K \neq 0$ is a complex number independent of τ. In the proof of (9) there are four cases to consider, corresponding to $q < \infty$ or $q = \infty$, and $q' < \infty$ and $q' = \infty$. If $q = q' = \infty$, then $c = 0$ and $d = \pm 1$. In this case the truth of (9) is obvious. Suppose that $q' < \infty$ and $q < \infty$. Then

$$\begin{aligned}(c\tau + d)^{-r}\sigma'(M\tau) &= (c\tau + d)^{-r}(M\tau - q')^{-r}\\ &= (c\tau + d)^{-r}(M\tau - Mq)^{-r}\\ &= (c\tau + d)^{-r}\left(\frac{\tau - q}{(c\tau + d)(cq + d)}\right)^{-r}\\ &= K(\tau - q)^{-r} = K\sigma(\tau),\end{aligned}$$

where $K \neq 0$ and K is independent of τ.

Suppose $q' < \infty$ and $q = \infty$. Then

$$\begin{aligned}(c\tau + d)^{-r}\sigma'(M\tau) &= (c\tau + d)^{-r}(M\tau - Mq)^{-r}\\ &= (c\tau + d)^{-r}\left(\frac{a\tau + b}{c\tau + d} - \frac{a}{c}\right)^{-r}\\ &= (c\tau + d)^{-r}\left(\frac{-1}{c(c\tau + d)}\right)^{-r} = K = K\sigma(\tau),\end{aligned}$$

where again $K \neq 0$ is independent of τ. If $q' = \infty$ and $q < \infty$, then

$$(c\tau + d)^{-r}\sigma'(M\tau) = (c\tau + d)^{-r}.$$

But $Mq = q' = \infty$ implies that $cq + d = 0$ or $q = -d/c$. Thus $(c\tau + d)^{-r} = [c(\tau - q)]^{-r} = K(\tau - q)^{-r} = K\sigma(\tau)$, with $K \neq 0$ and K independent of τ. This completes the proof of (9).

Consider the expansion

$$F(\tau) = \sigma'(\tau) \sum_{n=-\infty}^{\infty} a_n'(j)e^{2\pi i(n+\kappa')(A^{-1}M^{-1}\tau/\lambda)}$$

of $F(\tau)$ at q'. It follows that

$$F(M\tau) = \sigma'(M\tau) \sum_{n=-\infty}^{\infty} a_n'(j)e^{2\pi i(n+\kappa')A^{-1}\tau/\lambda}.$$

On the other hand, since $M \in \Gamma$,

$$F(M\tau) = v(M)(c\tau + d)^r F(\tau).$$

Thus

$$F(\tau) = \bar{v}(M)(c\tau + d)^{-r}\sigma'(M\tau) \sum_{n=-\infty}^{\infty} a_n'(j)e^{2\pi i(n+\kappa')A^{-1}\tau/\lambda},$$

and, applying (9), we obtain

$$F(\tau) = K'\sigma(\tau) \sum_{n=-\infty}^{\infty} a_n'(j) e^{2\pi i(n+\kappa')A^{-1}\tau/\lambda}, \tag{10}$$

with $K' \neq 0$ and independent of τ. From (10) it follows simply that

$$F(V\tau) = e^{2\pi i\kappa'} \frac{\sigma(V\tau)}{\sigma(\tau)} F(\tau),$$

while (5) implies that

$$F(V\tau) = e^{2\pi i\kappa} \frac{\sigma(V\tau)}{\sigma(\tau)} F(\tau).$$

Thus, since $F \not\equiv 0$, it follows that $\kappa = \kappa'$. Then (10) takes the form

$$F(\tau) = K'\sigma(\tau) \sum_{n=-\infty} a_n'(j) e^{2\pi i(n+\kappa)A^{-1}\tau/\lambda}.$$

A comparison with (5) yields $a_n(j) = K'a_n'(j)$, for all n, by the uniqueness of the Laurent expansion. Since K' is independent of n, the proof is complete.

Remark. If $q < \infty$, Lemma 5 implies that $\sigma(V\tau)/\sigma(\tau) = (\gamma\tau + \delta)^r$, where $V = \begin{pmatrix} * & * \\ \gamma & \delta \end{pmatrix}$. If $q = \infty$, a simple calculation shows that the same is true. Thus $F(V\tau) = e^{2\pi i\kappa}(\gamma\tau + \delta)^r F(\tau)$, which is consistent with (1).

4. DEFINITIONS OF MODULAR FUNCTION AND MODULAR FORM

We now suppose $F(\tau)$ is a function defined in $\mathscr{H}$ which satisfies the conditions in Theorem 4. Thus $F(\tau)$ has an expansion of the form (5) at each parabolic point q_j of the S.F.R. $\mathscr{R}$. We make the following

Definition. If only finitely many terms with $n < 0$ appear in the expansion (5) of $F(\tau)$ at q_j, we say that $F(\tau)$ is *meromorphic at* q_j. If $F(\tau)$ is meromorphic at q_j and the first nonzero $a_n(j)$ occurs for $n = -n_0 < 0$, we say that $F(\tau)$ has a *pole at* q_j *of order* $n_0 - \kappa_j$. If the first nonzero $a_n(j)$ occurs for $n = n_0 \geqslant 0$, we say that $F(\tau)$ is *regular at* q_j, with a *zero of order* $n_0 + \kappa_j$ *at* q_j.

Remarks. 1. The above terminology is justified by the fact that if $n + \kappa > 0$, then $e^{2\pi i(n+\kappa)A_j^{-1}\tau/\lambda_j} \to 0$ as $\tau \to q_j$, from within $\mathscr{R}$, and if $n + \kappa < 0$, then $|e^{2\pi i(n+\kappa)A_j^{-1}\tau/\lambda_j}| \to \infty$ as $\tau \to q_j$ from within $\mathscr{R}$. These facts follow, incidentally, from Lemma 1.

2. For the above definitions to be meaningful we must show that they depend only upon q_j and not upon its representation. Thus suppose $q_k =$

$A_k(\infty) = A_j(\infty) = q_j$, with $j \neq k$. Then by Remark 2 following the proof of Theorem 4, $a_n(j) = e^{2\pi i(n+\kappa_j)t/\lambda_j} a_n(k)$, for all n, where t is an integer. Thus $F(\tau)$ is meromorphic at q_j if and only if it is meromorphic at q_k, and the order of the zero or pole is the same at both q_j and q_k.

3. Suppose $\mathscr{R}$ and $\mathscr{R}'$ are two S.F.R.'s of Γ arranged again as in Proposition 3. Then $F(\tau)$ is meromorphic at q_j if and only if $F(\tau)$ is meromorphic at q_j', and furthermore the order of zero or pole is the same at both q_j and q_j'. These facts follow directly from Theorem 6.

We are finally in a position to define the important concept of a modular form.

Definition. Let r be a real number and let $v(M)$ be a multiplier system for the group Γ and the degree $-r$. A function $F(\tau)$ defined and meromorphic in $\mathscr{H}$ is said to be a *modular form of degree* $-r$, *with multiplier system* v, *with respect to* Γ if

(a) $F(\tau)$ satisfies (1), for all $M \in \Gamma$,

(b) there exists a S.F.R. $\mathscr{R}$ such that $F(\tau)$ has at most finitely many poles in $\bar{\mathscr{R}} \cap \mathscr{H}$ [hence $F(\tau)$ has expansions of the form (5) at the parabolic points of $\mathscr{R}$], and

(c) $F(\tau)$ is meromorphic at q_j, for each parabolic point q_j in $\mathscr{R}$.

Although conditions (b) and (c) are phrased in terms of a particular S.F.R. $\mathscr{R}$, we know from our previous discussion that if (b) and (c) are true for $F(\tau)$ in one S.F.R., then they hold in *all* S.F.R.'s.

We give three further important

Definitions. 1. If $F(\tau)$ is a modular form with respect to Γ we say that $F(\tau)$ is an *entire modular form* if

(a) $F(\tau)$ is regular in $\mathscr{H}$, and

(b) $F(\tau)$ is regular at all of the parabolic points of some S.F.R. $\mathscr{R}$.

2. If $F(\tau)$ is a modular form with respect to Γ such that $F(\tau)$ is regular in $\mathscr{H}$, and has a zero of positive order at all of the parabolic points of some S.F.R. $\mathscr{R}$, we say that $F(\tau)$ is a *cusp form.*

3. If $F(\tau)$ is a modular form on Γ with $r = 0$ and $v(M) = 1$, for all $M \in \Gamma$, then $F(\tau)$ is called a *modular function with respect to* Γ.

These definitions, too, are independent of the particular S.F.R. $\mathscr{R}$ which occurs.

5. SEVERAL IMPORTANT THEOREMS

In this section we prove several theorems which will serve as the basis of our applications of the theory of modular forms to number theory. We begin with the statement of

Theorem 7. If $f(\tau)$ is a modular function with respect to Γ such that $f(\tau)$ is regular in $\mathscr{H}$ and regular at all of the parabolic points of some

S.F.R. $\mathscr{R}$, then $f(\tau)$ is a constant. (That is, every entire modular function is constant.)

In the proof we shall use

Lemma 8. If $f(\tau)$ is as in Theorem 7, then for each parabolic point q_j in $\mathscr{R}$, $f(\tau) \to a_0(j)$ as $\tau \to q_j$ from within $\mathscr{R}$. Here $a_0(j)$ is the coefficient occurring in the expansion (5) of $f(\tau)$. Furthermore, $f(\tau)$ is bounded in $\mathscr{H}$.

Proof. Since $v(M) = 1$ for all $M \in \Gamma$, we have $\kappa_j = 0$ for $1 \leqslant j \leqslant \mu$. Also $r = 0$ and $f(\tau)$ is regular at all of the parabolic points q_j. Thus the expansion (5) of $f(\tau)$ at q_j has the form

(11) $$f(\tau) = \sum_{n=0}^{\infty} a_n(j) e^{2\pi i n (A_j^{-1}\tau)/\lambda_j},$$

valid in all of $\mathscr{H}$, since $f(\tau)$ is regular in $\mathscr{H}$. Now (11) is simply a power series in the variable $z = e^{2\pi i (A_j^{-1}\tau)/\lambda_j}$, convergent for $|z| < 1$. Thus as $z \to 0$, the series $\to a_0(j)$. By Lemma 1, as $\tau \to q_j$ from within $\mathscr{R}$, $(A_j^{-1}\tau)/\lambda_j \to \infty$ from within a vertical strip of the form $\operatorname{Im} w > 0$, $a_i < \operatorname{Re} w < b_i$. Consequently, as $\tau \to q_j$ from within $\mathscr{R}$, $z \to 0$, and $f(\tau) \to a_0(j)$.

It follows that $f(\tau)$ is bounded in the intersection of $\mathscr{R}$ and some open disc D_j centered at $q_j (1 \leqslant j \leqslant \mu)$. If $q_j = \infty$, then D_j is a half-plane of the form $y > y_j$, with $y_j > 0$. Let

(12) $$\mathscr{R}^* = \mathscr{R} - \bigcup_{j=1}^{\mu} D_j$$

(Figure 1). The closure of $\mathscr{R}^*$ is a compact subset of $\mathscr{H}$. Since $f(\tau)$ is regular in $\mathscr{H}$ it is continuous in the closure of $\mathscr{R}^*$ and therefore bounded in $\mathscr{R}^*$. Thus $f(\tau)$ is bounded in all of $\mathscr{R}$, and consequently bounded in $\bar{\mathscr{R}} \cap \mathscr{H}$. Since $f(M\tau) = f(\tau)$ for all $M \in \Gamma$, and $\mathscr{R}$ is a F.R. for Γ, it follows that $f(\tau)$ is bounded in $\mathscr{H}$.

Proof of Theorem 7. The tool used in the proof is the maximum modulus principle. Let $F(\tau) = \Pi_{j=1}^{\mu} \{f(\tau) - a_0(j)\}$. Then $F(\tau)$ is a modular function with respect to Γ. By Lemma 8, $F(\tau)$ is bounded in $\mathscr{H}$, and $F(\tau) \to 0$ as $\tau \to q_j$ from within $\mathscr{R}$, $1 \leqslant j \leqslant \mu$. Let C be the least upper bound of $|F(\tau)|$ in $\mathscr{H}$. We intend to show $C = 0$. By way of contradiction, suppose $C > 0$. Then there exists a set $\mathscr{R}^*$ of the form (12) such that in $\mathscr{R} - \mathscr{R}^*$, $|F(\tau)| < C/2$. It follows that the maximum modulus of $F(\tau)$ in $\bar{\mathscr{R}} \cap \mathscr{H}$, and hence in $\mathscr{H}$, occurs in $\mathscr{R}^* \subset \mathscr{H}$. This contradicts the maximum modulus principle, since $F(\tau)$ is regular in $\mathscr{H}$. Thus $C = 0$.

Now $F(\tau) \equiv 0$, so that, by continuity of $f(\tau)$ in $\mathscr{H}$, $f(\tau) \equiv a_0(j)$, for some fixed j. The proof is complete.

Theorem 7 can be rephrased as

Corollary 9. If $f(\tau)$ is a modular function with respect to Γ and $f(\tau)$ is bounded in $\mathscr{H}$, then $f(\tau)$ is a constant.

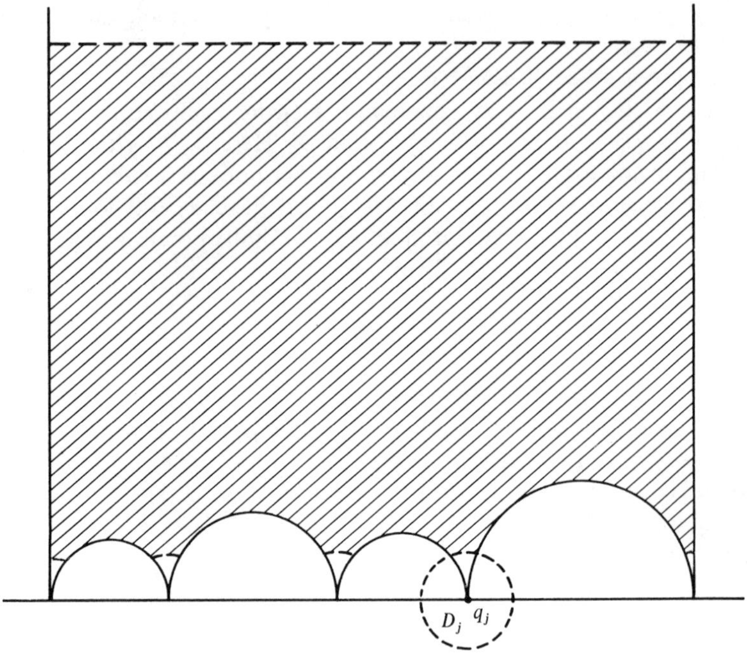

Figure 1. $\mathscr{R}$ is bounded by the vertical lines and solid circles. The shaded region is $\mathscr{R}^*$.

Proof. Since $f(\tau)$ is meromorphic and bounded in $\mathscr{H}$, it is actually regular in $\mathscr{H}$. Furthermore, the expansions (5) at q_j have the form

$$f(\tau) = \sum_{n=-n_0(j)}^{\infty} a_n(j) e^{2\pi i n (A_j^{-1}\tau)/\lambda_j},$$

valid in $\mathscr{H}$. If a term with $n < 0$ actually appeared in the expansion at a q_j, then $f(\tau)$ would not be bounded as $\tau \to q_j$ from within $\mathscr{R}$. Hence the expansions have the form (11) and the conclusion follows from Theorem 7.

Theorem 10. Suppose $F(\tau)$ is a cusp form of degree $-r$ with respect to Γ. Choose a S.F.R. $\mathscr{R}$, normalized so that $A_1 = I$. Thus $q_1 = \infty$. Put $\kappa = \kappa_1, \lambda = \lambda_1$, and $a_n = a_n(1)$. Then the expansion (5) at $q_1 = \infty$ has the form

$$F(\tau) = \sum_{n+\kappa>0} a_n e^{2\pi i (n+\kappa)\tau/\lambda}, \qquad \tau \in \mathscr{H}. \tag{13}$$

We conclude that for the a_n occurring in (13) $a_n = O(n^{r/2})$, as $n \to +\infty$; that is, $|a_n| \leqslant Cn^{r/2}$ for all $n \geqslant 1$, where C is a positive constant independent of n.

This theorem and the following lemma used in its proof are due to Hecke [*Werke* (Göttingen: Vandenhoech and Ruprecht, 1959), pp. 461–486, esp. p. 484].

Lemma 11. If $F(\tau)$ is a cusp form of degree $-r$ with respect to Γ, then $|F(x+iy)| \leqslant Cy^{-r/2}$, for all $\tau = x + iy \in \mathscr{H}$, where C is a positive constant independent of τ.

Proof. Put $\varphi(\tau) = \varphi(x+iy) = |y^{r/2}F(x+iy)|$ and let $M = \begin{pmatrix} * & * \\ \gamma & \delta \end{pmatrix} \in \Gamma$. Then $\operatorname{Im}(M\tau) = y/|\gamma\tau+\delta|^2$, and we have

$$\begin{aligned}\varphi(M\tau) &= |y^{r/2}(\gamma\tau+\delta)^{-r}F(M\tau)| = |y^{r/2}(\gamma\tau+\delta)^{-r}v(M)(\gamma\tau+\delta)^r F(\tau)| \\ &= |y^{r/2}F(\tau)| = \varphi(\tau);\end{aligned}$$

that is, $\varphi(\tau)$ is *invariant* with respect to Γ. We want to show that $\varphi(\tau)$ is bounded in $\mathscr{H}$; it is sufficient to show that $\varphi(\tau)$ is bounded in $\mathscr{R}$.

We shall show that as $\tau \to q_j$ from within $\mathscr{R}$, $\varphi(\tau) \to 0$, $1 \leqslant j \leqslant \mu$. This is similar to Lemma 8. The expansion (5) at q_j has the form

$$F(\tau) = \sigma_j(\tau) \sum_{n+\kappa_j>0} a_n(j)e^{2\pi i(n+\kappa_j)(A_j^{-1}\tau)/\lambda_j}, \qquad \tau \in \mathscr{H},$$

since $F(\tau)$ is a cusp form. Put $y' = \operatorname{Im}(A_j^{-1}\tau)$, and $A_j = \begin{pmatrix} a & * \\ c & * \end{pmatrix}$. Then $A_j^{-1} = \begin{pmatrix} * & * \\ -c & a \end{pmatrix}$ and $q_j = a/c$; $q_j < \infty$ if and only if $c \neq 0$. A short calculation shows that

$$y^{r/2} = (\operatorname{Im}\tau)^{r/2} = (y')^{r/2}|\sigma_j(\tau)|^{-1}\beta(c),$$

where

$$\beta(c) = \begin{cases} |c|^r & \text{if } c \neq 0, \\ 1 & \text{if } c = 0. \end{cases}$$

Thus for $\tau \in \mathscr{H}$,

$$\begin{aligned}\varphi(\tau) &= |y^{r/2}F(\tau)| = \beta(c)(y')^{r/2}\left|\sum_{n+\kappa_j>0} a_n(j)e^{2\pi i(n+\kappa_j)(A_j^{-1}\tau)/\lambda_j}\right| \\ &= \beta(c)(y')^{r/2}e^{-2\pi(n_0+\kappa_j)y'/\lambda_j}\left|\sum_{n\geqslant n_0} a_n(j)e^{2\pi i(n-n_0)A_j^{-1}\tau/\lambda_j}\right|,\end{aligned}$$

where $n_0 + \kappa_j > 0$. The infinite series on the right-hand side is a power series in the variable $z = e^{2\pi i(A_j^{-1}\tau)/\lambda_j}$, and thus $\to a_{n_0}(j)$ as $z \to 0$. By Lemma 1, $A_j^{-1}(\tau) \to \infty$ from within a vertical strip contained in $\mathscr{H}$, as $\tau \to q_j$ from within $\mathscr{R}$. Thus as $\tau \to q_j$ from within $\mathscr{R}$, $z \to 0$ and $y' \to +\infty$, so that $\varphi(\tau) \to 0$.

It follows that $\varphi(\tau)$ achieves its maximum inside a set $\mathscr{R}^*$ of the form (12). Since $\overline{\mathscr{R}^*}$ is a compact subset of $\mathscr{H}$ and $\varphi(\tau)$ is continuous in $\mathscr{H}$, $\varphi(\tau) \leqslant C$, $C > 0$, in $\overline{\mathscr{R}^*}$. Thus $\varphi(\tau) \leqslant C$ for all $\tau \in \mathscr{H}$. This completes the proof.

The same method of proof can be applied to give the following result. We leave the details for the reader to carry out.

Proposition 12. Suppose $F(\tau)$ is a modular form with respect to Γ, a subgroup of finite index in $\Gamma(1)$. Suppose also that q is a parabolic point in a S.F.R. $\mathscr{R}$ of Γ. Then if $F(\tau)$ has a zero of positive order at q, it follows that $F(\tau) \to 0$ as $\tau \to q$ from within $\mathscr{R}$.

Proof of Theorem 10. Let $z = x + iy \in \mathscr{H}$. Then

$$\begin{aligned}\int_z^{z+\lambda} & F(\zeta)e^{-2\pi i(n+\kappa)\zeta/\lambda}\, d\zeta \\ &= \int_z^{z+\lambda}\left(\sum_{m+\kappa>0} a_m e^{2\pi i(m+\kappa)\zeta/\lambda}\right)e^{-2\pi i(n+\kappa)\zeta/\lambda}\, d\zeta \\ &= \sum_{m+\kappa>0}\int_z^{z+\lambda} a_m e^{2\pi i(m-n)\zeta/\lambda}\, d\zeta = \lambda a_n,\end{aligned}$$

where the integral is taken along the horizontal path. The interchange of summation and integration is justified since the infinite sum is simply a power series in the variable $e^{2\pi i\zeta/\lambda}$. It follows that

$$a_n = \lambda^{-1}\int_z^{z+\lambda} F(\zeta)e^{-2\pi i(n+\kappa)\zeta/\lambda}\, d\zeta,$$

for all n, where z is any point in $\mathscr{H}$. By Lemma 11,

$$|a_n| \leqslant \lambda^{-1}Cy^{-r/2}e^{2\pi(n+\kappa)y/\lambda} = C_1y^{-r/2}e^{2\pi(n+\kappa)y/\lambda}. \tag{14}$$

Put $y = \operatorname{Im} z = 1/n$ and we obtain $|a_n| \leqslant C'n^{r/2}$, for all $n \geqslant 1$, where $C' = C_1e^{4\pi/\lambda}$ is independent of n. This completes the proof.

Corollary 13. If $r < 0$ in Theorem 10, that is, if $F(\tau)$ is a cusp form of positive degree, then $F(\tau)$ is identically 0.

Proof. By (14) we have $|a_n| \leqslant C_1y^{-r/2}e^{2\pi(n+\kappa)y/\lambda}$ for all n, where y is an arbitrary positive number. Letting $y \to 0+$, with $r < 0$, we obtain $a_n = 0$ for all n. Thus $F(\tau) \equiv 0$, and the proof is complete.

A stronger result is

Theorem 14. If $F(\tau)$ is an entire modular form of positive degree with respect to Γ, then $F(\tau)$ is identically zero.

Remark. This result is at the same time a generalization of Corollary 13 and an analogue of Theorem 7. It does not, however, include Theorem 7, as that theorem deals with entire modular *functions*, that is, entire forms of degree zero with multiplier system identically 1. Theorem 14 is well known and several proofs appear in the literature; the proof we give here appears to be somewhat different from those given previously.

Proof. We choose a S.F.R. $\mathscr{R}^*$ such that ∞ is *not* a cusp of $\mathscr{R}^*$. To see that this is possible observe that there exists an integer n (depending on Γ, of course) such that $W^n \in \Gamma$, where $W = \begin{pmatrix} 1 & 0 \\ 1 & 1 \end{pmatrix}$. If not $W, W^2, W^3, \ldots$ provide infinitely many cosets of $\Gamma(1)$ modulo Γ. Since Γ is of finite index in $\Gamma(1)$ this is impossible.

Suppose $\mathscr{R} = \bigcup_{j=1}^{\mu} A_j\{\mathscr{R}(\Gamma(1))\}$ is a S.F.R. for Γ. The cusps of $\mathscr{R}$ are the points $A_1(\infty), \ldots, A_\mu(\infty)$. If one of these, say $A_i(\infty)$, is the point ∞ itself, that is, if A_i is a translation, replace A_i by $B_i = W^n A_i$. If $A_i(\infty) \neq \infty$, put $B_i = A_i$. Then $B_1, \ldots, B_\mu$ are cosets for $\Gamma(1)$ modulo Γ in the sense that

$$\Gamma(1) = \bigcup_{j=1}^{\mu} \Gamma B_j,$$

and therefore

$$\mathscr{R}^* = \bigcup_{j=1}^{\mu} B_j\{R(\Gamma(1))\}$$

is a S.F.R. for Γ. The cusps of $\mathscr{R}^*$ are $B_1(\infty), \ldots, B_\mu(\infty)$ and none of these is ∞.

Write $q_j = B_j(\infty)$, $1 \leqslant j \leqslant \mu$. The expansion (5) of $F(\tau)$ at q_j has the form

$$F(\tau) = (\tau - q_j)^{-r} \sum_{n+\kappa_j \geqslant 0} a_n(j) e^{2\pi i(n+\kappa_j)B_j^{-1}\tau/\lambda_j}, \qquad \operatorname{Im} \tau > 0,$$

where $r < 0$. This follows since $F(\tau)$ is an entire modular form of positive degree and $q_j \neq \infty$.

As in the proof of Lemma 11, the function $\varphi(\tau) = \varphi(x + iy) = |y^{r/2} F(x + iy)|$ has the property $\varphi(V\tau) = \varphi(\tau)$, for all $V \in \Gamma$. We show that as $\tau \to q_j$ from within $\mathscr{R}^*$, $\varphi(\tau) \to 0$, $1 \leqslant j \leqslant \mu$. By Lemma 1, as $\tau \to q_j$ from within $\mathscr{R}^*$, $B_j^{-1}\tau \to \infty$ from within a vertical strip of the form $a_j < x < b_j$, $y \geqslant 0$, with $z = x + iy$. The calculation used in the proof of Lemma 11 shows that, for $\tau \in \mathscr{H}$,

$$\varphi(\tau) = |y^{r/2} F(\tau)| = |c|^r (y')^{r/2} \left| \sum_{n+\kappa_j \geqslant 0} a_n(j) e^{2\pi i(n+\kappa_j)B_j^{-1}\tau/\lambda_j} \right|,$$

where $y' = \operatorname{Im}(B_j^{-1}\tau)$, and $B_j = \begin{pmatrix} * & * \\ c & * \end{pmatrix}$. Since $y' \to +\infty$ as $\tau \to q_j$ from within $\mathscr{R}^*$ and $r < 0$, we conclude that $\varphi(\tau) \to 0$ as $\tau \to q_j$ from within $\mathscr{R}^*$.

As in the proof of Lemma 11, it follows that there exists $C > 0$, independent of τ, such that $\varphi(\tau) \leqslant C$ for all $\tau \in \mathscr{H}$. Therefore, $|F(\tau)| \leqslant Cy^{-r/2}$, $\tau \in \mathscr{H}$.

We proceed as in the proof of Theorem 10 to find that if we consider the expansion of $F(\tau)$ at ∞,

$$F(\tau) = \sum_{m+\kappa>0} a_m e^{2\pi i(m+\kappa)\tau/\lambda}, \qquad \operatorname{Im}\tau > 0,$$

we conclude that $|a_n| \leqslant Cy^{-r/2}e^{2\pi(n+\kappa)y/\lambda}$ for $n = 0, 1, 2,\ldots$ and arbitrary $y > 0$. Since $r < 0$, we get $a_n = 0$, for all n, by letting $y \to 0+$. Hence $F(\tau) = 0$ for all $\tau \in \mathscr{H}$ and the theorem is proved.

Remark. This method when applied to modular forms of degree zero, and in particular to modular functions, shows only that $|a_n| \leqslant C$, for all n. As we have seen in Theorem 7 the analogue of Theorem 14 is valid if $r = 0$ and $v = 1$ on Γ. If $r = 0$, but $v \not\equiv 1$ on Γ, the result if still true, but we are not in a position to establish it here.

We close this chapter with a result which is simple but of some importance in our later developments.

Theorem 15. Let $\Gamma_2 \subset \Gamma_1 \subset \Gamma(1)$, where Γ_1 and Γ_2 are subgroups of finite index in $\Gamma(1)$. Suppose $\mu = [\Gamma_1 : \Gamma_2]$ and we have the coset decomposition $\Gamma_1 = \bigcup_{i=1}^{\mu} \Gamma_2 A_i$. Let $g(\tau)$ be an invariant function with respect to Γ_2 in the sense that $g(M\tau) = g(\tau)$ for all $M \in \Gamma_2$. Then if $F(x_1,\ldots,x_\mu)$ is symmetric in its μ variables, the function $f(\tau) = F(g(A_1\tau),\ldots,g(A_\mu\tau))$ is an invariant function with respect to Γ_1.

Proof. Let $M \in \Gamma_1$. Then $A_iM = M_iA_i'$, where $M_i \in \Gamma_2$ and $A_i' \in \{A_1,\ldots,A_\mu\}$. We assert that as A_i runs through a complete set of right coset representatives, then so does A_i'. For suppose $A_iM = M_iA_i'$, $A_jM = M_jA_j'$, with $A_i' = A_j'$. Then

$$\begin{aligned}(A_iM)(A_jM)^{-1} &= A_iA_j^{-1} = (M_iA_i')(M_jA_j')^{-1}\\ &= M_iA_i'A_j'^{-1}M_j^{-1} = M_iM_j^{-1} \in \Gamma_2.\end{aligned}$$

Since $A_1,\ldots,A_\mu$ is a complete set of distinct right coset representatives, it follows that $A_i = A_j$ and $i = j$.

We next observe that $g(A_i\tau)$ is independent of the choice of A_i from the coset Γ_2A_i. For if $A_i^* \in \Gamma_2A_i$, then $A_i^* = MA_i$, with $M \in \Gamma_2$. It follows that $g(A_i^*\tau) = g(MA_i\tau) = g(A_i\tau)$, by the invariance of $g(\tau)$ with respect to Γ_2. Now let $M \in \Gamma_1$. We have

$$\begin{aligned}f(M\tau) &= F(g(A_1M\tau),\ldots,g(A_\mu M\tau)) = F(g(M_1A_1'\tau),\ldots,g(M_\mu A_\mu'\tau))\\ &= F(g(A_1'\tau),\ldots,g(A_\mu'\tau)) = f(\tau),\end{aligned}$$

by the invariance of $g(\tau)$ with respect to Γ_2 and the symmetry of the function F. The proof is complete.

Remark. If $g(\tau)$ is a modular function with respect to Γ_2 and $F(x_1,...,x_\mu)$ is a symmetrical polynomial, then $f(\tau)$ is a modular function with respect to Γ_1. The proof is left for the reader.

Chapter 3

THE MODULAR FORMS $\eta(\tau)$ AND $\vartheta(\tau)$

1. THE FUNCTION $\eta(\tau)$

For $\tau \in \mathscr{H}$ we define $\eta(\tau)$ by

$$\eta(\tau) = e^{\pi i\tau/12} \prod_{m=1}^{\infty} (1 - e^{2\pi i m\tau}).$$

Simple considerations of complex function theory show that $\eta(\tau)$ is regular in $\mathscr{H}$ and $\eta(\tau) \neq 0$ in $\mathscr{H}$.

Definition. Let n be a positive integer. Then $p(n)$ is the number of partitions of n into positive integers. Here the parts are not necessarily distinct, but the order is disregarded. That is, $2 + 1$ and $1 + 2$ are regarded as the same partition of 3.

Examples. 1. $p(3) = 3$. For 3 can be partitioned as $3, 2 + 1, 1 + 1 + 1$.

2. $p(7) = 15$. For 7 can be partitioned as $7, 6 + 1, 5 + 2, 4 + 3, 5 + 1 + 1, 4 + 2 + 1, 3 + 3 + 1, 3 + 2 + 2, 4 + 1 + 1 + 1, 3 + 2 + 1 + 1, 2 + 2 + 2 + 1, 3 + 1 + 1 + 1 + 1, 2 + 2 + 1 + 1 + 1, 2 + 1 + 1 + 1 + 1 + 1, 1 + 1 + 1 + 1 + 1 + 1 + 1$.

Let us also define $p(0) = 1$. We can observe without too much difficulty that, at least in a formal sense,

$$\prod_{n=1}^{\infty} (1 - x^n)^{-1} = \sum_{m=0}^{\infty} p(m)x^m. \tag{1}$$

In order to verify (1) formally, we recall that $(1 - x^n)^{-1} = \Sigma_{j=0}^{\infty} x^{nj}$, so that

$$\prod_{n=1}^{\infty} (1 - x^n)^{-1} = \prod_{n=1}^{\infty} \left(\sum_{j=0}^{\infty} x^{nj} \right). \tag{2}$$

Now, given a nonnegative integer m, the coefficient of x^m on the right-hand side of (2) is the number of ways of writing m in the form $m = 1j_1 + 2j_2 + 3j_3 + \cdots$, where each j_i is a nonnegative integer. On the other hand, given a partition of m, we may rewrite the partition in the form

$$m = 1j_1 + 2j_2 + 3j_3 + \cdots,$$

by simply collecting all the 1's in the partition together, then all the 2's, and so forth. This method of rewriting gives a 1–1 correspondence between the partitions of m and the number of occurrences of x^m on the right-hand side of (2). This proves (1) in a formal sense.

We now prove (1) rigorously.

Proposition 1. For $|x| < 1$ both the left-hand and right-hand sides of (1) converge and they are equal.

Proof. It is clear that $\Pi_{n=1}^{\infty}(1 - x^n)$ converges to an analytic function with no zeros in $|x| < 1$, since $\Sigma_{n=1}^{\infty} |x|^n$ converges for $|x| < 1$. Hence $\Pi_{n=1}^{\infty} (1 - x^n)^{-1}$ converges in $|x| < 1$ to an analytic function with no zeros.

We next show that the power series on the right-hand side of (1) converges, for $0 \leqslant x < 1$, to the left-hand side of (1). Let $0 \leqslant x < 1$ and let N be a positive integer. Consider

$$\prod_{n=1}^{N} (1 - x^n)^{-1} = \prod_{n=1}^{N} \left(\sum_{j=0}^{\infty} x^{nj} \right).$$

If $m \leqslant N$, then every expression of the form $m = 1j_1 + 2j_2 + 3j_3 + \cdots$ must stop with the Nth term; that is, the expression reduces to $m = 1j_1 + 2j_2 + \cdots + Nj_N$. Thus the coefficient of x^m in $\Pi_{n=1}^{N} (\Sigma_{j=0}^{\infty} x^{nj})$ is $p(m)$, if $0 \leqslant m \leqslant N$, and we can write

$$(3) \qquad \prod_{n=1}^{N} (1 - x^n)^{-1} = \sum_{m=0}^{N} p(m)x^m + \sum_{m=N+1}^{\infty} a_m x^m,$$

where $0 \leqslant a_m \leqslant p(m)$ for $m \geqslant N + 1$. From (3) we conclude that, for $0 \leqslant x < 1$,

$$\sum_{m=0}^{N} p(m)x^m \leqslant \prod_{n=1}^{N} (1 - x^n)^{-1} \leqslant \prod_{n=1}^{\infty} (1 - x^n)^{-1}.$$

The last inequality holds since each factor $(1 - x^n)^{-1}$ is $\geqslant 1$. Since $p(m)x^m \geqslant 0$, $\Sigma_{m=0}^{N} p(m)x^m$ is a monotone increasing function of N. Thus for $0 \leqslant x < 1$, $\Sigma_{m=0}^{\infty} p(m)x^m$ converges and $\Sigma_{m=0}^{\infty} p(m)x^m \leqslant \Pi_{n=1}^{\infty} (1 - x^n)^{-1}$. On the other hand, it also follows from (3) that $\Sigma_{m=0}^{\infty} p(m)x^m \geqslant \Pi_{n=1}^{N} (1 - x^n)^{-1}$, for each positive integer N. Thus letting $N \to +\infty$ we conclude that

$$\sum_{m=0}^{\infty} p(m)x^m \geqslant \prod_{n=1}^{\infty} (1 - x^n)^{-1}.$$

Thus for $0 \leqslant x < 1$, $\Pi_{n=1}^{\infty} (1 - x^n)^{-1} = \Sigma_{m=0}^{\infty} p(m)x^m$.

Now suppose only that $|x| < 1$. The left-hand side of (1) is regular for $|x| < 1$. The right-hand side of (1) is a power series which converges absolutely in $|x| < 1$. For $\Sigma_{m=0}^{\infty} |p(m)x^m| = \Sigma_{m=0}^{\infty} p(m)|x|^m$, and $0 \leqslant |x| < 1$. Thus the right-hand side is also regular in $|x| < 1$. By the identity theorem for analytic functions (1) holds for all x such that $|x| < 1$.

If we rephrase Proposition 1 in terms of the function $\eta(\tau)$ we get

Corollary 2. For $\tau \in \mathscr{H}$,

$$\eta^{-1}(\tau) = \sum_{m=0}^{\infty} p(m) e^{2\pi i[m-(1/24)]\tau}$$
$$= \sum_{m=-1}^{\infty} p(m+1) e^{2\pi i[m+(23/24)]\tau}.$$

It is clear from Corollary 2 that the function $\eta(\tau)$ will be important in the study of the number-theoretic function $p(n)$.

2. SEVERAL FAMOUS IDENTITIES

The results of this section are several famous and important identities, involving $\eta(\tau)$, which are due to Euler and Jacobi. The basic one, from which the others will follow is

Theorem 3 (Jacobi's Identity). Suppose x and z are complex numbers such that $|x| < 1$ and $z \neq 0$. Then

$$(4) \qquad \prod_{n=0}^{\infty} (1 - x^{2n+2})(1 + x^{2n+1}z)(1 + x^{2n+1}z^{-1}) = \sum_{m=-\infty}^{\infty} x^{m^2} z^m.$$

The proof we give of (4) is a recent one due to G. E. Andrews [*Proc. Am. Math. Soc.*, *16* (1965), pp. 333–334]. We begin with a preliminary result.

Lemma 4 (Euler). (a) For $|x| < 1$ and any complex z,

$$\prod_{n=0}^{\infty} (1 + x^n z) = \sum_{m=0}^{\infty} \frac{x^{m(m-1)/2} z^m}{(1-x)(1-x^2)\cdots(1-x^m)}.$$

(b) For $|x| < 1$, $|z| < 1$,

$$\prod_{n=0}^{\infty} (1 + x^n z)^{-1} = \sum_{m=0}^{\infty} \frac{(-1)^m z^m}{(1-x)(1-x^2)\cdots(1-x^m)}.$$

Proof. (a) Let $f(x,z) = \Pi_{n=0}^{\infty} (1 + x^n z)$. This infinite product converges absolutely for $|x| < 1$ and any z. Write $f(x,z) = \Sigma_{m=0}^{\infty} a_m(x) z^m$, a power-series expansion for $f(x,z)$ about $z = 0$ with fixed x such that $|x| < 1$. It is clear that this power-series expansion is valid for all complex z. From the definition of $f(x,z)$ it follows immediately that $f(x,z) = (1 + z) f(x,zx)$. Hence

$$\sum_{m=0}^{\infty} a_m(x) z^m = \sum_{m=0}^{\infty} a_m(x) x^m z^m + \sum_{m=0}^{\infty} a_m(x) x^m z^{m+1}.$$

It follows that for $m > 0$,

$$a_m(x) = a_m(x)x^m + a_{m-1}(x)x^{m-1},$$

or

$$a_m(x) = a_{m-1}(x)x^{m-1}(1 - x^m)^{-1}.$$

Since $a_0(x) = 1$, we have, by induction,

$$a_m(x) = \frac{x^{(m-1)+(m-2)+\cdots+1}}{(1-x)(1-x^2)\cdots(1-x^m)} = \frac{x^{m(m-1)/2}}{(1-x)(1-x^2)\cdots(1-x^m)}.$$

The result follows.

(b) In this case put $g(x,z) = \prod_{n=0}^{\infty}(1 + x^n z)^{-1}$. As long as $|x| < 1$ and $|z| < 1$ this product converges absolutely. With fixed x such that $|x| < 1$ it represents a function of z analytic for $|z| < 1$. Thus the power-series expansion $g(x,z) = \sum_{m=0}^{\infty} b_m(x)z^m$ is valid for $|z| < 1$. From the definition of $g(x,z)$ it is clear that $g(x,zx) = (1 + z)g(x,z)$, so that

$$\sum_{m=0}^{\infty} b_m(x)x^m z^m = \sum_{m=0}^{\infty} b_m(x)z^m + \sum_{m=0}^{\infty} b_m(x)z^{m+1}.$$

Hence for $m > 0$, $b_m(x)x^m = b_m(x) + b_{m-1}(x)$, or

$$b_m(x) = \frac{-b_{m-1}(x)}{(1 - x^m)}.$$

Since $b_0(x) = 1$ we conclude that

$$b_m(x) = \frac{(-1)^m}{(1-x)(1-x^2)\cdots(1-x^m)},$$

and the proof is complete.

Proof of Theorem 3. Assume that $|x| < 1$ and $z \neq 0$. By Lemma 4(a),

$$\begin{aligned}
\prod_{n=0}^{\infty}(1 + x^{2n+1}z) &= \prod_{n=0}^{\infty}(1 + (x^2)^n(xz)) \\
&= \sum_{m=0}^{\infty} \frac{x^{2m(m-1)/2}x^m z^m}{(1-x^2)\cdots(1-x^{2m})} \\
&= \sum_{m=0}^{\infty} \frac{x^{m^2}z^m}{(1-x^2)\cdots(1-x^{2m})} \\
&= \sum_{m=0}^{\infty} x^{m^2}z^m \left\{ \frac{\prod_{j=0}^{\infty}(1 - x^{2m+2+2j})}{\prod_{j=0}^{\infty}(1 - x^{2j+2})} \right\} \\
&= \prod_{j=0}^{\infty}(1 - x^{2j+2})^{-1} \sum_{m=0}^{\infty} x^{m^2}z^m \prod_{j=0}^{\infty}(1 - x^{2m+2+2j}).
\end{aligned}$$

Note that if $m < 0$, $\Pi_{j=0}^{\infty}(1 - x^{2m+2+2j}) \equiv 0$. Thus the above can be written as

$$\prod_{j=0}^{\infty}(1 - x^{2j+2})^{-1} \sum_{m=-\infty}^{\infty} x^{m^2} z^m \prod_{j=0}^{\infty}(1 - x^{2m+2+2j}).$$

We again apply Lemma 4(a) to get

$$\begin{aligned}\prod_{j=0}^{\infty}(1 - x^{2m+2+2j}) &= \prod_{j=0}^{\infty}(1 + (x^2)^j(-x^{2+2m})) \\ &= \sum_{k=0}^{\infty} \frac{x^{2k(k-1)/2}(-x^{2+2m})^k}{(1 - x^2)\cdots(1 - x^{2k})} \\ &= \sum_{k=0}^{\infty} \frac{(-1)^k x^{k^2+k+2mk}}{(1 - x^2)\cdots(1 - x^{2k})}.\end{aligned}$$

Combining this with the earlier calculation we obtain

$$(5)\quad \prod_{n=0}^{\infty}(1 + x^{2n+1}z)$$
$$= \prod_{j=0}^{\infty}(1 - x^{2j+2})^{-1} \sum_{m=-\infty}^{\infty} \sum_{k=0}^{\infty} \frac{(-1)^k z^m x^{m^2+k^2+2mk+k}}{(1 - x^2)\cdots(1 - x^{2k})},$$

valid for $|x| < 1$ and any complex z. Although the double sum on the right-hand side of (5) converges for $|x| < 1$ and all complex z it converges *absolutely* only for $|x| < 1$ and $|z| > |x|$. With these restrictions we may then interchange the order of summation to obtain

$$\begin{aligned}&\sum_{k=0}^{\infty} \frac{(-1)^k x^k}{(1 - x^2)\cdots(1 - x^{2k})} \sum_{m=-\infty}^{\infty} x^{(m+k)^2} z^m \\ &= \sum_{k=0}^{\infty} \frac{(-1)^k (xz^{-1})^k}{(1 - x^2)\cdots(1 - x^{2k})} \sum_{m=0}^{\infty} x^{(m+k)^2} z^{m+k} \\ &= \sum_{k=0}^{\infty} \frac{(-1)^k (xz^{-1})^k}{(1 - x^2)\cdots(1 - x^{2k})} \sum_{m=-\infty}^{\infty} x^{m^2} z^m.\end{aligned}$$

Thus for $|x| < 1$ and $|z| > |x|$, (5) becomes

$$(6)\quad \prod_{n=0}^{\infty}(1 + x^{2n+1}z)$$
$$= \left(\sum_{m=-\infty}^{\infty} x^{m^2} z^m\right) \prod_{j=0}^{\infty}(1 - x^{2j+2})^{-1} \sum_{k=0}^{\infty} \frac{(-1)^k (xz^{-1})^k}{(1 - x^2)\cdots(1 - x^{2k})}.$$

Next we apply Lemma 4(b) to the sum on the right-hand side of (6) to get

$$\sum_{k=0}^{\infty} \frac{(-1)^k(xz^{-1})^k}{(1-x^2)\cdots(1-x^{2k})} = \prod_{n=0}^{\infty} (1 + (x^2)^n(xz^{-1}))^{-1},$$

valid for $|x| < 1$ and $|x| < |z|$. Thus, under these conditions on x and z, we have

$$\prod_{n=0}^{\infty} (1 + x^{2n+1}z) = \left(\sum_{m=-\infty}^{\infty} x^{m^2}z^m\right) \prod_{j=0}^{\infty} (1 - x^{2j+2})^{-1} \prod_{n=0}^{\infty} (1 + x^{2n+1}z^{-1})^{-1},$$

or

$$(7) \quad \sum_{m=-\infty}^{\infty} x^{m^2}z^m = \prod_{n=0}^{\infty} (1 - x^{2n+2})(1 + x^{2n+1}z)(1 + x^{2n+1}z^{-1}).$$

With a fixed x such that $|x| < 1$, both sides of (7) are analytic functions of z for $z \neq 0$. Thus by the identity theorem for analytic functions, (7) holds for $|x| < 1$ and $z \neq 0$. The proof is now complete.

Corollary 5 (Euler's Identity). For $|x| < 1$ we have

$$\prod_{n=1}^{\infty} (1 - x^n) = \sum_{m=-\infty}^{\infty} (-1)^m x^{m(3m+1)/2}.$$

Proof. In Theorem 3 replace x by $x^{3/2}$ and z by $-x^{1/2}$. For $0 < |x| < 1$ we get

$$\prod_{n=0}^{\infty} (1 - x^{3n+3})(1 - x^{(2n+1)3/2}x^{1/2})(1 - x^{(2n+1)3/2}x^{-1/2})$$
$$= \sum_{m=-\infty}^{\infty} (x^{3/2})^{m^2}(-x^{1/2})^m.$$

A short calculation yields

$$\prod_{n=0}^{\infty} (1 - x^{3n+3})(1 - x^{3n+2})(1 - x^{3n+1}) = \sum_{m=-\infty}^{\infty} (-1)^m x^{m(3m+1)/2},$$

or

$$\prod_{n=1}^{\infty} (1 - x^n) = \sum_{m=-\infty}^{\infty} (-1)^m x^{m(3m+1)/2},$$

for $0 < |x| < 1$.

Obviously the equality holds for $x = 0$.

Corollary 6 (Jacobi's Identity). For $|x| < 1$ we have

$$\prod_{n=1}^{\infty} (1 - x^n)^3 = \sum_{m=0}^{\infty} (-1)^m(2m + 1)x^{m(m+1)/2}.$$

Proof. One is tempted here simply to replace x by $x^{1/2}$ and z by $-x^{1/2}$ in Theorem 3, but unfortunately this reduces to 0 on both sides of the identity. Thus we must be slightly subtler.

In Theorem 3 replace x by $x^{1/2}$ and z by $x^{1/2}(-1+\varepsilon)$, with $0<\varepsilon<1$. For $|x|<1$ and $0<\varepsilon<1$, we get

$$\prod_{n=1}^{\infty}(1-x^n)(1+x^n(-1+\varepsilon))(1+x^{n-1}(-1+\varepsilon)^{-1}) = \sum_{m=-\infty}^{\infty} x^{m(m+1)/2}(-1+\varepsilon)^m,$$

or

$$\text{(8)}\qquad \frac{1}{\varepsilon-1}\prod_{n=1}^{\infty}(1-x^n)(1+x^n(-1+\varepsilon))(1+x^n(-1+\varepsilon)^{-1}) = \varepsilon^{-1}\sum_{m=-\infty}^{\infty} x^{m(m+1)/2}(-1+\varepsilon)^m.$$

With fixed x such that $|x|<1$ denote the left-hand side of (8) by $f(\varepsilon)$ and the right-hand side by $g(\varepsilon)$. The form of $f(\varepsilon)$ shows that it is an analytic function of ε for $\varepsilon \neq 1$. For our present argument the important point is that $f(\varepsilon)$ is right-hand continuous at $\varepsilon=0$. Thus, by (8),

$$\text{(9)}\qquad \lim_{\varepsilon\to 0+} g(\varepsilon) = \lim_{\varepsilon\to 0+} f(\varepsilon) = f(0) = -\prod_{n=1}^{\infty}(1-x^n)^3.$$

We next examine $\lim_{\varepsilon\to 0+} g(\varepsilon)$. Write

$$(-1+\varepsilon)^m = (-1)^m(1-m\varepsilon-\rho), \qquad \text{with } \rho = 1-(1-\varepsilon)^m - m\varepsilon.$$

By Taylor's theorem,

$$(1-\varepsilon)^m = 1 - m\varepsilon + \frac{m(m-1)}{2}t^{m-2}\varepsilon^2,$$

where $1-\varepsilon \leqslant t \leqslant 1$, so that, for all integers m,

$$|\rho| = \frac{m(m-1)}{2}t^{m-2}\varepsilon^2 < \tfrac{1}{2}(|m|+1)^2\varepsilon^2.$$

Thus

$$g(\varepsilon) = \varepsilon^{-1}\sum_{m=-\infty}^{\infty} x^{m(m+1)/2}(-1+\varepsilon)^m$$
$$= \varepsilon^{-1}\sum_{m=-\infty}^{\infty} x^{m(m+1)/2}(-1)^m + \varepsilon^{-1}\sum_{m=-\infty}^{\infty}(-1)^{m+1}(m\varepsilon)x^{m(m+1)/2} + R,$$

where $R = \varepsilon^{-1} \Sigma_{m=-\infty}^{\infty} x^{m(m+1)/2}(-1)^{m+1}\rho$. Now

$$|R| \leqslant \varepsilon^{-1} \sum_{m=-\infty}^{\infty} |x|^{m(m+1)/2}|\rho| < \frac{\varepsilon}{2} \sum_{m=-\infty}^{\infty} (|m| + 1)^2|x|^{m(m+1)/2} = K\varepsilon,$$

where K depends on x but not on ε. Thus for $|x| < 1$ $\lim_{\varepsilon\to 0+} R = 0$. On the other hand, a simple calculation shows that $\Sigma_{m=-\infty}^{\infty} x^{m(m+1)/2}(-1)^m = 0$, for $|x| < 1$. Hence $\lim_{\varepsilon\to 0+} g(\varepsilon) = \Sigma_{m=-\infty}^{\infty} (-1)^{m+1} m x^{m(m+1)/2}$, and a comparison with (9) yields $\Pi_{n=1}^{\infty} (1 - x^n)^3 = \Sigma_{m=-\infty}^{\infty} (-1)^m m x^{m(m+1)/2}$. Combining the terms for m and $-m-1$, with $m \geqslant 0$, we find that the last sum becomes $\Sigma_{m=0}^{\infty} (-1)^m(2m+1)x^{m(m+1)/2}$, and the proof is complete.

3. TRANSFORMATION FORMULAS FOR $\eta(\tau)$

In this section we show that $\eta(\tau)$ is a cusp form of degree $-\frac{1}{2}$ with respect to the full modular group, $\Gamma(1)$. Along the way we obtain several other interesting and important results. The first of these is

Theorem 7 (Poisson's Sum Formula). Suppose $f(t)$ is a continuous function for $-\infty < t < \infty$. Suppose also that $\Sigma_{m=-\infty}^{\infty} f(t+m)$ is uniformly convergent for $0 \leqslant t \leqslant 1$. Then

$$\sum_{m=-\infty}^{\infty} f(t+m) = \sum_{n=-\infty}^{\infty} e^{-2\pi int} \int_{-\infty}^{\infty} f(x)e^{2\pi inx}\,dx,$$

for all t such that the right-hand side converges.

Proof. Put $\varphi(t) = \Sigma_{m=-\infty}^{\infty} f(t+m)$, which, by our assumptions is continuous for all real t and periodic, with period 1, in t. We consider the Fourier series expansion of $\varphi(t)$. With

$$a_n = \int_0^1 \varphi(x)e^{2\pi inx}\,dx,$$

we know that $\Sigma_{n=-\infty}^{\infty} a_n e^{-2\pi int}$ is Cesaro-summable to $\varphi(t)$, for all t. This follows from Fejer's theorem, since $\varphi(t)$ is continuous for all t. [See, for example, E. C. Titchmarsh, *The Theory of Functions* (2nd ed., rev.; New York: Oxford University Press, 1952), pp. 412–414.] In particular, by an elementary summability principle it converges to $\varphi(t)$ whenever it actually converges.

Since $\Sigma_{m=-\infty}^{\infty} f(t+m)$ converges uniformly for $0 \leqslant t \leqslant 1$,

$$\begin{aligned} a_n &= \int_0^1 \varphi(x)e^{2\pi inx}\,dx = \int_0^1 \sum_{m=-\infty}^{\infty} f(x+m)e^{2\pi inx}\,dx \\ &= \sum_{m=-\infty}^{\infty} \int_0^1 f(x+m)e^{2\pi inx}\,dx = \sum_{m=-\infty}^{\infty} \int_m^{m+1} f(x)e^{2\pi inx}\,dx \\ &= \int_{-\infty}^{\infty} f(x)e^{2\pi inx}\,dx. \end{aligned}$$

It follows that

$$\varphi(t) = \sum_{n=-\infty}^{\infty} a_n e^{-2\pi int} = \sum_{n=-\infty}^{\infty} e^{-2\pi int} \int_{-\infty}^{\infty} f(x) e^{2\pi inx}\, dx,$$

for any t such that the right-hand side converges.

As an important consequence of the Poisson sum formula we derive

Theorem 8 (Theta Transformation Formula). For all complex z and complex t with $\operatorname{Re} t > 0$, we have

$$\sum_{n=-\infty}^{\infty} e^{-\pi t(n+z)^2} = \frac{1}{\sqrt{t}} \sum_{n=-\infty}^{\infty} e^{-\pi n^2/t + 2\pi inz},$$

where the branch of $\sqrt{t}$ is determined according to the convention that $|\arg t| < \pi/2$.

Proof. Put $g(z) = e^{-\pi tz^2}$, for complex z. Clearly $g(z)$ is continuous for all real z and $\sum_{n=-\infty}^{\infty} g(z+n) = \sum_{n=-\infty}^{\infty} e^{-\pi t(z+n)^2}$ converges uniformly for $0 \leqslant z \leqslant 1$, since $\operatorname{Re} t > 0$. Thus by Theorem 7,

$$\sum_{n=-\infty}^{\infty} e^{-\pi t(z+n)^2} = \sum_{n=-\infty}^{\infty} e^{-2\pi inz} \int_{-\infty}^{\infty} e^{-\pi tx^2 + 2\pi inx}\, dx, \tag{10}$$

provided the series on the right-hand side converges. We shall show that the right-hand side converges absolutely for all real z under the restriction $t > 0$. With $t > 0$ we make the substitution $y = \sqrt{t}\,x$ in the integral and obtain

$$\int_{-\infty}^{\infty} e^{-\pi tx^2 + 2\pi inx}\, dx = \frac{1}{\sqrt{t}} \int_{-\infty}^{\infty} e^{-\pi y^2 + 2\pi iny/\sqrt{t}}\, dy = e^{-\pi n^2/t} \frac{\gamma}{\sqrt{t}},$$

where we have put

$$\gamma = \int_{-\infty}^{\infty} e^{-\pi(y - in/\sqrt{t})^2}\, dy = \int_{-\infty - in/\sqrt{t}}^{\infty - in/\sqrt{t}} e^{-\pi y^2}\, dy.$$

Let $\mathscr{P}$ be the rectangular path with vertices $\pm T$, $\pm T - (in/\sqrt{t})$ (see Figure 2). By Cauchy's theorem $\int_{\mathscr{P}} e^{-\pi y^2}\, dy = 0$. This, together with the fact that the integrand tends uniformly to 0 on the vertical parts of $\mathscr{P}$ as $T \to \infty$, implies that $\gamma = \int_{-\infty}^{\infty} e^{-\pi y^2}\, dy$, a constant independent of n and t. The right-hand side of (10) becomes

$$\frac{\gamma}{\sqrt{t}} \sum_{n=-\infty}^{\infty} e^{-2\pi inz} e^{-\pi n^2/t},$$

a series which obviously converges absolutely for all real z. Thus for $t > 0$

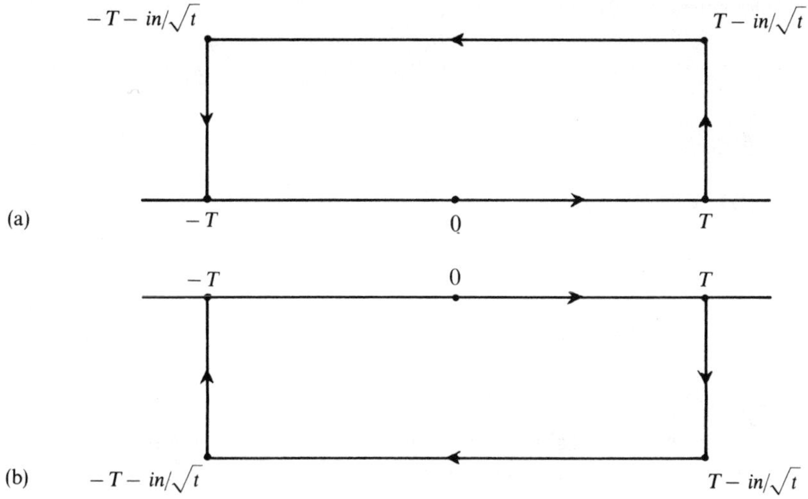

Figure 2. (a) The path $\mathscr{P}$, with $n < 0$. (b) The path $\mathscr{P}$, with $n > 0$.

and real z,

$$\sum_{n=-\infty}^{\infty} e^{-\pi t(z+n)^2} = \frac{\gamma}{\sqrt{t}} \sum_{n=-\infty}^{\infty} e^{-\pi n^2/t + 2\pi i n z}.$$

Putting $z = 0$ and $t = 1$ we find that $\gamma = 1$. Thus we have

(11) $$\sum_{n=-\infty}^{\infty} e^{-\pi t(z+n)^2} = \frac{1}{\sqrt{t}} \sum_{n=-\infty}^{\infty} e^{-\pi n^2/t + 2\pi i n z},$$

valid for $t > 0$ and real z.

With fixed $t > 0$ both sides of (11) are analytic functions of z for all complex z. Therefore, by the indentity theorem for analytic functions (11) holds for all complex z with $t > 0$. On the other hand, with fixed complex z both sides of (11) are analytic functions of t for $\operatorname{Re} t > 0$. The identity theorem then implies that (11) is valid for $\operatorname{Re} t > 0$ and z complex. This completes the proof.

From Theorem 8 we derive the important transformation formula for $\eta(\tau)$ contained in the next result.

Theorem 9. For $\tau \in \mathscr{H}$, we have

$$\eta\left(-\frac{1}{\tau}\right) = (-i)^{1/2}\tau^{1/2}\eta(\tau),$$

where the roots are calculated according to the convention established in Chapter 2, Section 1; that is, $-\pi \leqslant \arg z < \pi$, for a complex number z.

Proof. By definition,

$$\eta(\tau) = e^{\pi i\tau/12} \prod_{m=1}^{\infty} (1 - e^{2\pi i m\tau}), \qquad \tau \in \mathscr{H}.$$

By Corollary 5, we have

$$\begin{aligned}
\eta(\tau) &= e^{\pi i\tau/12} \sum_{n=-\infty}^{\infty} (-1)^n e^{\pi i\tau n(3n+1)} = e^{\pi i\tau/12} \sum_{n=-\infty}^{\infty} (-1)^n e^{\pi i\tau n(3n-1)} \\
&= e^{\pi i\tau/12} \sum_{n=-\infty}^{\infty} e^{3\pi i\tau\{n-(1/6)(1-1/\tau)\}^2} e^{-(\pi i\tau/12)(1-1/\tau)^2} \\
&= e^{-\pi i/12\tau} e^{\pi i/6} \sum_{n=-\infty}^{\infty} e^{3\pi i\tau\{n-(1/6)(1-1/\tau)\}^2},
\end{aligned}$$

after a number of straightforward algebraic manipulations in the exponent. We next apply Theorem 8, with $t = -3i\tau$ and $z = -\frac{1}{6}(1 - 1/\tau)$. This yields

$$\eta(\tau) = e^{-\pi i/12\tau} e^{\pi i/6} \frac{1}{\sqrt{-3i\tau}} \sum_{n=-\infty}^{\infty} e^{-\pi i n^2/3\tau - \pi i n(1-1/\tau)/3}.$$

Breaking up the summation on n according to residue class modulo 3, we obtain

$$\sum_{n=-\infty}^{\infty} = g_0(\tau) + g_1(\tau) + g_2(\tau),$$

where

$$g_k(\tau) = \sum_{\mu=-\infty}^{\infty} e^{-\pi i(3\mu+k)^2/3\tau} e^{-\pi i(3\mu+k)(1-1/\tau)/3},$$

for $k = 0, 1, 2$. Now

$$g_0(\tau) = \sum_{\mu=-\infty}^{\infty} (-1)^\mu e^{-(\pi i/\tau)(3\mu^2-\mu)},$$

$$\begin{aligned}
g_1(\tau) &= \sum_{\mu=-\infty}^{\infty} e^{-\pi i(3\mu+1)^2/3\tau} e^{-\pi i(3\mu+1)(1-1/\tau)/3} \\
&= e^{-\pi i/3} \sum_{\mu=-\infty}^{\infty} (-1)^\mu e^{-\pi i(3\mu^2-\mu)/\tau} = e^{-\pi i/3} g_0(\tau),
\end{aligned}$$

and

$$\begin{aligned}
g_2(\tau) &= \sum_{\mu=-\infty}^{\infty} e^{-(\pi i/3\tau)(3\mu+2)^2} e^{-(\pi i/3)(3\mu+2)(1-1/\tau)} \\
&= e^{-2\pi i/3\tau} e^{-2\pi i/3} \sum_{\mu=-\infty}^{\infty} (-1)^\mu e^{-(3\pi i/\tau)\mu(\mu+1)}.
\end{aligned}$$

As we have already observed in the proof of Corollary 6,

$$\sum_{\mu=-\infty}^{\infty} (-1)^{\mu} e^{-(3\pi i/\tau)\mu(\mu+1)} = 0, \qquad \text{for } \tau \in \mathscr{H}.$$

Hence $g_2(\tau) = 0$ for $\tau \in \mathscr{H}$, and we find that

$$\eta(\tau) = e^{-\pi i/12\tau} e^{\pi i/6} \frac{1}{\sqrt{-3i\tau}} (g_0(\tau) + e^{-\pi i/3} g_0(\tau))$$

$$= e^{-\pi i/12\tau} \frac{1}{\sqrt{3}} \frac{1}{\sqrt{-i\tau}} g_0(\tau)(e^{\pi i/6} + e^{-\pi i/6})$$

$$= \frac{1}{\sqrt{-i\tau}} e^{-\pi i/12\tau} \sum_{\mu=-\infty}^{\infty} (-1)^{\mu} e^{-(\pi i/\tau)(3\mu^2-\mu)} = \frac{1}{\sqrt{-i\tau}} \eta(-1/\tau).$$

Here again we have used Corollary 5. It now follows that

$$\eta(-1/\tau) = \sqrt{-i\tau}\,\eta(\tau) = (-i)^{1/2}\tau^{1/2}\eta(\tau),$$

as a check of the branches of the square roots reveals.

Theorem 10. $\eta(\tau)$ is a modular form of degree $-\frac{1}{2}$ on the full modular group $\Gamma(1)$. In fact, $\eta(\tau)$ is a cusp form of degree $-\frac{1}{2}$ on $\Gamma(1)$. Furthermore, if we denote the multiplier system for $\eta(\tau)$ by v_η we have $v_\eta(S) = e^{\pi i/12}$, $v_\eta(T) = (-i)^{1/2} = e^{-\pi i/4}$. $v_\eta(M)$ is a 24th root of unity, for all $M \in \Gamma(1)$.

Proof. We want to prove that

$$\eta(M\tau) = v_\eta(M)(c\tau + d)^{1/2}\eta(\tau), \tag{12}$$

for all $M = \begin{pmatrix} * & * \\ c & d \end{pmatrix} \in \Gamma(1)$, where $|v_\eta(M)| = 1$.

In Theorem 7 we proved that $\eta(T\tau) = \eta(-1/\tau) = (-i)^{1/2}\tau^{1/2}\eta(\tau)$, while it is obvious from the definition of $\eta(\tau)$ that $\eta(S\tau) = \eta(\tau + 1) = e^{\pi i/12}\eta(\tau)$. Thus we already know that (12) holds for $M = T$ and $M = S$. The essential point here is that, by Corollary 3 of Chapter 1, S and T *generate* $\Gamma(1)$. We shall be able to prove (12) for all $M \in \Gamma(1)$ by knowing that it holds for a set of generators of $\Gamma(1)$.

Any $M \in \Gamma(1)$ can be written as a word of the form

$$M = T^{\varepsilon_1} S^{\alpha_1} T \cdots T S^{\alpha_n} T^{\varepsilon_2}$$

where $\varepsilon_1 = 0, 1, 2$, or 3; $\varepsilon_2 = 0$ or 1; and $\alpha_1, \ldots, \alpha_n$ are rational integers. Here we are thinking of $\Gamma(1)$ as a group of *matrices*. Thus it is necessary to allow ε_1 to be 2 or 3 in order to include the elements $-I$ and $-T$. Define the *length* of M to be $\varepsilon_1 + |\alpha_1| + \cdots + |\alpha_n| + \varepsilon_2 + n - 1$. Note that the length of M is not unique because $T^2 = (ST)^3 = -I$ in the matrix group

$\Gamma(1)$. This fact, although not pleasing, has no real effect on our proof, which proceeds by induction on the length.

Suppose M has length 0. Then $M = I$ and (12) obviously holds with $v_\eta(I) = 1$. Suppose (12) is known for all $M \in \Gamma(1)$ such that M can be expressed as a word of length $k-1$, $k \geqslant 1$. Suppose that $M' \in \Gamma(1)$ and M' can be expressed as a word of length k. Then we have either $M' = MS$, or $M' = MT$, or $M' = MS^{-1}$, where $M = \begin{pmatrix} * & * \\ c & d \end{pmatrix}$ has length $k-1$. If $M' = MS$, then

$$\eta(M'\tau) = \eta(MS\tau) = v_\eta(M)(cS\tau + d)^{1/2}\eta(S\tau) = v_\eta(M)e^{\pi i/12}(c\tau + d + c)^{1/2}\eta(\tau)$$
$$= v_\eta(M')(c\tau + d + c)^{1/2}\eta(\tau),$$

where we have put $v_\eta(M') = v_\eta(M)e^{\pi i/12}$. This shows that (12) holds for M' since $M' = MS = \begin{pmatrix} * & * \\ c & c+d \end{pmatrix}$ and $|v_\eta(M')| = 1$. Since M' has many different expressions as a word in S and T we should comment on the uniqueness of $v_\eta(M')$. That we get the same value for $v_\eta(M')$ regardless of how we express M' as a word follows from the fact that $v_\eta(M') = \eta(M'\tau)(c\tau + d + c)^{-1/2}/\eta(\tau)$. This same comment is relevant for the cases $M' = MT$ and $M' = MS^{-1}$ treated below, but will not be repeated there.

If $M' = MT$, then

$$\eta(M'\tau) = \eta(MT\tau)$$
$$= v_\eta(M)(cT\tau + d)^{1/2}\eta(T\tau) = v_\eta(M)(-i)^{1/2}(-c/\tau + d)^{1/2}\tau^{1/2}\eta(\tau).$$

Clearly

$$(-c/\tau + d)^{1/2}\tau^{1/2} = e^{\pi i\lambda(d,c)}(d\tau - c)^{1/2},$$

where $\lambda(d,c)$ is a rational integer independent of $\tau \in \mathscr{H}$. Therefore, $\eta(M'\tau) = v_\eta(M')(d\tau - c)^{1/2}\eta(\tau)$, with $v_\eta(M') = v_\eta(M)(-i)^{1/2}e^{\pi i\lambda(d,c)}$. Since $|v_\eta(M')| = 1$ and $M' = MT = \begin{pmatrix} * & * \\ d & -c \end{pmatrix}$, it follows that (12) holds for M'.

If $M' = MS^{-1}$, then $\eta(M'\tau) = \eta(MS^{-1}\tau) = v_\eta(M)(cS^{-1}\tau + d)^{1/2}\eta(S^{-1}\tau) = v_\eta(M)e^{-\pi i/12}(c\tau + d - c)^{1/2}\eta(\tau)$. Putting $v_\eta(M') = v_\eta(M)e^{-\pi i/12}$, we find that $|v_\eta(M')| = 1$. Since also $M' = MS^{-1} = \begin{pmatrix} * & * \\ c & d-c \end{pmatrix}$, (12) holds for M'. By induction, then, (12) holds for all $M \in \Gamma(1)$.

Also $\eta(\tau)$ is analytic in $\mathscr{H}$ and, by Corollary 5, $\eta(\tau)$ has the following Fourier expansion at ∞:

$$\eta(\tau) = e^{\pi i\tau/12} \sum_{m=-\infty}^{\infty} (-1)^m e^{\pi i m(3m+1)\tau}.$$

With $\kappa = \frac{1}{24}$ we see that only exponents of the form $2\pi i(n + \kappa)$, $n + \kappa > 0$, occur. Since ∞ is the only parabolic cusp of $\Gamma(1)$, it follows that $\eta(\tau)$ is a cusp form of degree $-\frac{1}{2}$ on $\Gamma(1)$.

4. THE FUNCTION $\vartheta(\tau)$

For $\tau \in \mathscr{H}$, we define the function $\vartheta(\tau)$ by

$$\vartheta(\tau) = \sum_{n=-\infty}^{\infty} e^{\pi i n^2\tau} = 1 + 2\sum_{n=1}^{\infty} e^{\pi i n^2 \tau}.$$

It is clear from elementary complex function theory that $\vartheta(\tau)$ is regular in $\mathscr{H}$.

Remark. The number-theoretic significance of $\vartheta(\tau)$ is contained in the fact that for s a nonnegative integer,

$$\vartheta^s(\tau) = 1 + \sum_{n=1}^{\infty} r_s(n) e^{\pi i n\tau}, \tag{13}$$

where $r_s(n)$ is defined to be the number of ways of representing n as a sum of s squares. Here the parts are not necessarily distinct and the order is significant. Thus, for example, $r_2(25) = 12$, since

$$25 = (\pm 5)^2 + 0^2 = 0^2 + (\pm 5)^2 = (\pm 4)^2 + (\pm 3)^2 = (\pm 3)^2 + (\pm 4)^2.$$

With this definition of $r_s(n)$, (13) follows immediately from the absolute convergence of the series for $\vartheta(\tau)$.

It is not at all obvious from the definition that $\vartheta(\tau) \neq 0$ for $\tau \in \mathscr{H}$. However, this is true, and in fact we have the following result.

Theorem 11. For $\tau \in \mathscr{H}$, we have

$$\vartheta(\tau) = \prod_{n=1}^{\infty} (1 - e^{2\pi i n\tau})(1 + e^{(2n-1)\pi i\tau})^2. \tag{14}$$

Thus, in particular, $\vartheta(\tau) \neq 0$ for $\tau \in \mathscr{H}$.

Proof. Apply Theorem 3 (Jacobi's identity) with $x = e^{\pi i\tau}$ and $z = 1$. This yields

$$\vartheta(\tau) = \sum_{n=-\infty}^{\infty} e^{\pi i n^2\tau} = \prod_{n=0}^{\infty} (1 - e^{(2n+2)\pi i\tau})(1 + e^{(2n+1)\pi i\tau})^2$$

$$= \prod_{n=1}^{\infty} (1 - e^{2\pi i n\tau})(1 + e^{(2n-1)\pi i\tau})^2,$$

the desired result. That $\vartheta(\tau) \neq 0$ for $\tau \in \mathscr{H}$ follows from the absolute convergence of the infinite product on the right-hand side of (14).

The identity (14) not only shows that $\vartheta(\tau) \neq 0$ for $\tau \in \mathscr{H}$, but also establishes an important connection between $\vartheta(\tau)$ and $\eta(\tau)$, the function discussed earlier in this chapter. This connection is contained in the next theorem.

Theorem 12. For $\tau \in \mathscr{H}$, we have

$$\vartheta(\tau) = \eta^2\left(\frac{\tau + 1}{2}\right) \bigg/ \eta(\tau + 1). \tag{15}$$

Proof. By definition,

$$\eta(\tau) = e^{\pi i\tau/12} \prod_{m=1}^{\infty} (1 - e^{2\pi i m\tau}), \qquad \tau \in \mathscr{H}.$$

Therefore,

$$\eta^2\left(\frac{\tau + 1}{2}\right) = e^{\pi i\tau/12} e^{\pi i/12} \prod_{m=1}^{\infty} (1 - e^{\pi i m(\tau + 1)})^2,$$

and

$$\eta(\tau + 1) = e^{\pi i\tau/12} e^{\pi i/12} \prod_{m=1}^{\infty} (1 - e^{2\pi i m\tau}).$$

Thus, forming the quotient, we obtain

$$\eta^2\left(\frac{\tau + 1}{2}\right) \bigg/ \eta(\tau + 1) = \prod_{m=1}^{\infty} (1 - (-1)^m e^{\pi i m\tau})^2 (1 - e^{2\pi i m\tau})^{-1}$$

$$= \frac{\prod_{m=1}^{\infty} (1 - e^{2\pi i m\tau})^2 (1 + e^{(2m-1)\pi i\tau})^2}{\prod_{m=1}^{\infty} (1 - e^{2\pi i m\tau})}$$

$$= \prod_{m=1}^{\infty} (1 - e^{2\pi i m\tau})(1 + e^{(2m-1)\pi i\tau})^2.$$

This last product, however, is $\vartheta(\tau)$, by Theorem 11. This completes the proof.

From the transformation properties of $\eta(\tau)$ (Theorem 10) we can derive similar properties of $\vartheta(\tau)$.

Theorem 13. The function $\vartheta(\tau)$ is an entire modular form of degree $-\frac{1}{2}$ on the subgroup Γ_ϑ of $\Gamma(1)$. If we denote the multiplier system for $\vartheta(\tau)$ by v_ϑ, then $v_\vartheta(S^2) = 1$ and $v_\vartheta(T) = (i)^{-1/2} = e^{-\pi i/4}$. Finally, the expansion of $\vartheta(\tau)$ at ∞ has the form $\vartheta(\tau) = 1 + 2\sum_{n=1}^{\infty} e^{\pi i n^2\tau}$, and the expansion at -1

has the form

$$\vartheta(\tau) = (\tau + 1)^{-1/2} \sum_{n=0}^{\infty} b_n e^{2\pi i[n+(1/8)](-1/(\tau+1))}, \qquad b_n \neq 0.$$

Remarks. 1. By definition of Γ_ϑ it is generated by S^2 and T.

2. In Chapter 2 we observed that Γ_ϑ has a S.F.R. in which ∞ and -1 are the parabolic points.

Proof. Let $M = \begin{pmatrix} a & b \\ c & d \end{pmatrix} \in \Gamma_\vartheta$. Then by Corollary 4 of Chapter 1, $a \equiv d \pmod 2$ and $b \equiv c \pmod 2$. Put

$$M_1 = \begin{pmatrix} a + c & \dfrac{b-c}{2} + \dfrac{d-a}{2} \\ 2c & d - c \end{pmatrix}, \qquad M_2 = \begin{pmatrix} a + c & b + d \\ c & d \end{pmatrix},$$

and observe that both M_1 and M_2 are in $\Gamma(1)$. Then by Theorem 12

$$(16) \quad \vartheta(M\tau) = \eta^2\left(\frac{M\tau + 1}{2}\right) \Big/ \eta(M\tau + 1) = \eta^2\left(M_1\left(\frac{\tau+1}{2}\right)\right) \Big/ \eta(M_2\tau).$$

By Theorem 10,

$$\eta\left(M_1\left(\frac{\tau+1}{2}\right)\right) = v_\eta(M_1)\left(2c\left(\frac{\tau+1}{2}\right) + d - c\right)^{1/2} \eta\left(\frac{\tau+1}{2}\right)$$
$$= v_\eta(M_1)(c\tau + d)^{1/2}\eta\left(\frac{\tau+1}{2}\right),$$

and

$$\eta(M_2\tau) = v_\eta(M_2)(c\tau + d)^{1/2}\eta(\tau)$$
$$= v_\eta(M_2)e^{-\pi i/12}(c\tau + d)^{1/2}\eta(\tau + 1).$$

Therefore, (16) becomes

$$\vartheta(M\tau) = v_\eta^2(M_1)v_\eta(M_2)^{-1}e^{\pi i/12}(c\tau + d)^{1/2} \frac{\eta^2\left(\dfrac{\tau+1}{2}\right)}{\eta(\tau+1)}$$
$$= v_\eta^2(M_1)v_\eta(M_2)^{-1}e^{\pi i/12}(c\tau + d)^{1/2}\vartheta(\tau).$$

Since the first three factors on the right-hand side have absolute value 1, $\vartheta(\tau)$ satisfies the appropriate functional equation for a modular form of degree $-\frac{1}{2}$ on Γ_ϑ. The multiplier system v_ϑ can be expressed as

$$(17) \qquad v_\vartheta(M) = v_\eta^2(M_1)v_\eta(M_2)^{-1}e^{\pi i/12}.$$

That $v_\vartheta(S^2) = 1$ follows immediately from the original definition of $\vartheta(\tau)$ as the Fourier series

$$\vartheta(\tau) = 1 + 2 \sum_{n=1}^{\infty} e^{\pi i n^2 \tau}.$$

This series is, of course, also the expansion of $\vartheta(\tau)$ at ∞.

We wish to calculate $v_\vartheta(T)$. Observe that $\vartheta(T\tau) = v_\vartheta(T)\tau^{1/2}\vartheta(\tau)$ and put $\tau = i$. We obtain $\vartheta(i) = v_\vartheta(T)(i)^{1/2}\vartheta(i)$. Since $\vartheta(i) \neq 0$ by Theorem 11, we find that

$$v_\vartheta(T) = (i)^{-1/2} = e^{-\pi i/4}.$$

There remains only the calculation of the expansion of $\vartheta(\tau)$ at the cusp -1. To carry this out consider $\vartheta(-1 - 1/\tau)$. By Theorem 12,

$$\vartheta(-1 - 1/\tau) = \eta^2\left(\frac{-1/\tau}{2}\right) \Big/ \eta(-1/\tau) = \eta^2(-1/2\tau)/\eta(-1/\tau).$$

An application of Theorem 10 and a simple calculation together show that

$$\vartheta(-1 - 1/\tau) = (-i)^{1/2} 2\tau^{1/2} \eta^2(2\tau)/\eta(\tau).$$

Applying the original definition of $\eta(\tau)$ as an infinite product, we obtain

$$\vartheta(-1 - 1/\tau) = 2(-i)^{1/2}\tau^{1/2}e^{\pi i\tau/4} \prod_{m=1}^{\infty} (1 - e^{4\pi i m\tau})^2(1 - e^{2\pi i m\tau})^{-1}$$

$$= \tau^{1/2} \sum_{n=0}^{\infty} a_n e^{2\pi i[n + (1/8)]\tau}, \qquad a_0 = 2(-i)^{1/2} \neq 0.$$

Replacing τ by $-1/(\tau + 1)$, we obtain

$$\vartheta(\tau) = (\tau + 1)^{-1/2} \sum_{n=0}^{\infty} i a_n e^{2\pi i(n + 1/8)[-1/(\tau + 1)]},$$

the desired expansion of $\vartheta(\tau)$ at -1. This completes the proof.

Chapter 4

THE MULTIPLIER SYSTEMS v_η AND v_ϑ

1. PRELIMINARIES

This chapter is devoted to the proofs of exact formulas for v_η and v_ϑ in terms of the Jacobi symbol of elementary number theory. These exact formulas are totally indispensable in the applications we give in Chapters 5 to 7 of the theory of $\eta(\tau)$ and $\vartheta(\tau)$ to the number-theoretic functions $p(n)$ and $r_s(n)$. The formula we give for v_η was first found by Petersson [*Abhandl. Deut. Akad. Wiss. Berlin, 2* (1954), 59 pp.]. The first exact formula for v_η was given by Rademacher in 1931 [*J. Reine Angew. Math., 167* (1931), pp. 312–336] in terms of the Dedekind sums.

The approach we adopt here is to simply state the formulas and verify them by means of an inductive proof, the induction proceeding on the length of a group element as a word in the generators. (This is merely a computationally more complicated version of the simple idea which already appears in the proof of Theorem 10 of Chapter 3.) This approach has the obvious drawback of not giving any indication of how the formulas were discovered. On the other hand, it has the advantage of relative brevity and simplicity. Alternatively, one could derive the Rademacher formulas and either use these directly in the applications or reduce them to the formulas we give here by calculating the Dedekind sums in terms of the Jacobi symbol. In the latter approach one would use the results of Rademacher and Whiteman [*Am. J. Math., 63* (1941), pp. 377–407].

We begin with a definition of the Jacobi symbol and then state, without proof, some of its most important properties.

Definition. (a) If p is a prime number and a an integer, we call a a *quadratic* *residue* modulo p if there exists an integer x such that $x^2 \equiv a \pmod{p}$. If p is an odd prime and a is an integer such that $(a,p) = 1$ we define *Legendre's symbol* $\left(\frac{a}{p}\right)$ by

$$\left(\frac{a}{p}\right) = \begin{cases} 1 & \text{if } a \text{ is a quadratic residue mod } p, \\ -1 & \text{otherwise.} \end{cases}$$

(b) Suppose b is a positive odd integer and a is an integer such that $(a,b) = 1$. If $b > 1$ write $b = p_1 \cdots p_s$, where the p_i are primes, not necessarily distinct. Then *Jacobi's symbol* $\left(\frac{a}{b}\right)$ is defined by $\left(\frac{a}{b}\right) = \left(\frac{a}{p_1}\right) \cdots \left(\frac{a}{p_s}\right)$, where the factors on the right-hand side are Legendre's symbols. Furthermore, we put $\left(\frac{a}{1}\right) = 1$. Note that if b is itself an odd prime, then Jacobi's symbol reduces to Legendre's symbol. Thus there is no danger in the customary practice of using the same notation for both concepts.

The following result will be needed in our proofs of the formulas for v_η and v_ϑ. Although nontrivial [for example, (f) is the law of quadratic reciprocity and (g) is a generalization], the proof belongs to the domain of "elementary" number theory and is readily available in several places. Thus we do not include a proof, but merely refer the reader to E. Landau, *Elementary Number Theory*, part 1 (2nd ed., New York: Chelsea Publishing Co., Inc., 1966), pp. 65–69.

Lemma 1. (a) If m is an odd positive integer, $(n,m) = 1$, and $n' \equiv n \pmod{m}$, then $\left(\frac{n}{m}\right) = \left(\frac{n'}{m}\right)$.

(b) If m and m' are odd positive integers and $(n,m) = 1 = (n,m')$, then $\left(\frac{n}{m}\right)\left(\frac{n}{m'}\right) = \left(\frac{n}{mm'}\right)$.

(c) If m is an odd positive integer and $(n,m) = 1 = (n',m)$, then $\left(\frac{n}{m}\right)\left(\frac{n'}{m}\right) = \left(\frac{nn'}{m}\right)$.

(d) If m is an odd positive integer, then $\left(\frac{-1}{m}\right) = (-1)^{\frac{m-1}{2}}$.

(e) If m is an odd positive integer, then $\left(\frac{2}{m}\right) = (-1)^{\frac{m^2-1}{8}}$.

(f) (Law of Quadratic Reciprocity). If m and n are odd positive integers such that $(m,n) = 1$, then

$$\left(\frac{n}{m}\right)\left(\frac{m}{n}\right) = (-1)^{\frac{n-1}{2}\frac{m-1}{2}}.$$

(g) If m and n are odd and $(m,n) = 1$, then

$$\left(\frac{n}{|m|}\right)\left(\frac{m}{|n|}\right) = \begin{cases} -(-1)^{\frac{n-1}{2}\frac{m-1}{2}} & \text{if } n < 0,\ m < 0, \\ (-1)^{\frac{n-1}{2}\frac{m-1}{2}} & \text{otherwise.} \end{cases}$$

We need one further definition before we can state the main theorems.

Definition. Suppose c and d are integers such that $(c,d) = 1$ and d is odd. We then put $\left(\frac{c}{d}\right)^* = \left(\frac{c}{|d|}\right)$ and $\left(\frac{c}{d}\right)_* = \left(\frac{c}{|d|}\right)(-1)^{\frac{\operatorname{sign} c - 1}{2}\frac{\operatorname{sign} d - 1}{2}}$, provided $c \neq 0$. We complete the definition by putting $\left(\frac{0}{\pm 1}\right)^* = 1$, $\left(\frac{0}{1}\right)_* = 1$, and $\left(\frac{0}{-1}\right)_* = -1$. As usual, if x is a nonzero real number, then $\operatorname{sign} x = x/|x|$.

We are now in a position to state the main results of this chapter.

Theorem 2. The multiplier system v_η of the modular form $\eta(\tau)$ is given by the following formula: for each $M = \begin{pmatrix} a & b \\ c & d \end{pmatrix} \in \Gamma(1)$,

$$v_\eta(M) = \begin{cases} \left(\dfrac{d}{c}\right)^* \exp\left\{\dfrac{\pi i}{12}[(a + d)c - bd(c^2 - 1) - 3c]\right\} & \text{if } c \text{ is odd,} \\ \left(\dfrac{c}{d}\right)_* \exp\left\{\dfrac{\pi i}{12}[(a + d)c - bd(c^2 - 1) + 3d - 3 - 3cd]\right\} & \text{if } c \text{ is even.} \end{cases}$$

Theorem 3. The multiplier system v_ϑ of the modular form $\vartheta(\tau)$ is given by the formula: For each $M = \begin{pmatrix} a & b \\ c & d \end{pmatrix} \in \Gamma_\vartheta$,

$$v_\vartheta(M) = \begin{cases} \left(\dfrac{d}{c}\right)^* e^{-\pi i c/4} & \text{if } b \equiv c \equiv 1,\ a \equiv d \equiv 0 \pmod 2, \\ \left(\dfrac{c}{d}\right)_* e^{\pi i (d-1)/4} & \text{if } a \equiv d \equiv 1,\ b \equiv c \equiv 0 \pmod 2. \end{cases}$$

2. PROOF OF THEOREM 2

The reader is referred to the proof of Theorem 10 of Chapter 3 for the definition of the *length* of an element of $\Gamma(1)$ as a word in the generators S and T. Our proof will proceed by induction on the length. Of great importance in the proof is the "consistency condition" satisfied by v_η. For $M_1 = \begin{pmatrix} * & * \\ c_1 & d_1 \end{pmatrix}$, and $M_2 = \begin{pmatrix} * & * \\ c_2 & d_2 \end{pmatrix} \in \Gamma(1)$, we have

$$(1)\quad v_\eta(M_1M_2)(c_3\tau + d_3)^{1/2} = v_\eta(M_1)v_\eta(M_2)(c_1M_2\tau + d_1)^{1/2}(c_2\tau + d_2)^{1/2},$$

where $M_1M_2 = \begin{pmatrix} * & * \\ c_3 & d_3 \end{pmatrix}$. This is simply the general consistency condition (3) of Chapter 2 for the particular case $v = v_\eta$.

The only element of $\Gamma(1)$ that can be expressed as a word of length 0 is of course $I = \begin{pmatrix} 1 & 0 \\ 0 & 1 \end{pmatrix}$. But $v_\eta(I) = 1$ and the formula gives, in this case, $\left(\frac{0}{1}\right)_* e^{\pi i/(3-3)/12} = 1$. Thus, in this case, the proposed formula is correct. Suppose the formula is correct for all elements of $\Gamma(1)$ that can be expressed as words of length $n - 1$, where n is a fixed integer $\geqslant 1$. Suppose also $M' \in \Gamma(1)$ such that M' can be expressed as a word of length n. Then either $M' = MS$, $M' = MS^{-1}$, or $M' = MT$, where $M = \begin{pmatrix} a & b \\ c & d \end{pmatrix}$ has length $n - 1$. The proof is broken up into three cases corresponding to these three possible expressions for M'.

Case 1. $M' = MS = \begin{pmatrix} a & b \\ c & d \end{pmatrix}\begin{pmatrix} 1 & 1 \\ 0 & 1 \end{pmatrix} = \begin{pmatrix} a & a+b \\ c & c+d \end{pmatrix}.$

(a) Suppose c is odd. Then of course $c \neq 0$ and the induction hypothesis applied to M yields

$$v_\eta(M) = \left(\frac{d}{c}\right)^* \exp\left\{\frac{\pi i}{12}[(a + d)c - bd(c^2 - 1) - 3c]\right\}.$$

On the other hand, we know that $v_\eta(S) = e^{\pi i/12}$. By (1),

$$v_\eta(M')(c\tau + c + d)^{1/2} = v_\eta(M)v_\eta(S)(c\tau + c + d)^{1/2},$$

or

$$v_\eta(M') = v_\eta(M)v_\eta(S) = \left(\frac{d}{c}\right)^* \exp\left\{\frac{\pi i}{12}[(a + d)c - bd(c^2 - 1) - 3c + 1]\right\}.$$

The proposed formula for $v_\eta(M')$ gives

$$\left(\frac{c+d}{c}\right)^* \exp\left\{\frac{\pi i}{12}[(a+c+d)c-(a+b)(c+d)(c^2-1)-3c]\right\}$$

$$=\left(\frac{d}{c}\right)^* \exp\left\{\frac{\pi i}{12}[(a+d)c-bd(c^2-1)-3c]\right\}\exp\left(\frac{\pi i}{12}E\right),$$

where $E = c^2 - (ac + bc + ad)(c^2 - 1)$. Here we have used the fact that $\left(\frac{c+d}{c}\right)^* = \left(\frac{d}{c}\right)^*$, which follows from the definition of $\left(\frac{d}{c}\right)^*$ and Lemma 1(a). Thus, to verify the proposed formula for $v_\eta(M')$ it is sufficient to prove that $E \equiv 1 \pmod{24}$. This can be accomplished by a simple calculation using the fact that $ad - bc = 1$. The details are left to the reader.

(b) Suppose c is even. By the induction hypothesis

$$v_\eta(M) = \left(\frac{c}{d}\right)_* \exp\left\{\frac{\pi i}{12}[(a+d)c-bd(c^2-1)+3d-3-3cd]\right\},$$

so that

$$v_\eta(M') = v_\eta(M)v_\eta(S) = \left(\frac{c}{d}\right)_* \exp\left\{\frac{\pi i}{12}[(a+d)c-bd(c^2-1)+3d-2-3cd]\right\}.$$

The proposed formula for $v_\eta(M')$, with which we must compare this, is

$$\left(\frac{c}{c+d}\right)_* \exp\left\{\frac{\pi i}{12}[(a+c+d)c-(a+b)(c+d)(c^2-1)+3(c+d)-3-3c(c+d)]\right\}.$$

Since c is even, $c + d$ is odd. If $c = 0$, then $a = d = \pm 1$ and in this case $v_\eta(M') = \left(\frac{0}{\pm 1}\right)_* \exp\left\{\frac{\pi i}{12}(bd+3d-2)\right\}$, while the proposed formula gives

$$\left(\frac{0}{\pm 1}\right)_* \exp\left\{\frac{\pi i}{12}[(a+b)d+3d-3]\right\} = \left(\frac{0}{\pm 1}\right)_* \exp\left\{\frac{\pi i}{12}(bd+3d-2)\right\}.$$

Thus the proposed formula is correct if $c = 0$. Assume therefore that $c \neq 0$ and write $c = 2^\alpha c_1$, where c_1 is odd and $\alpha \geqslant 1$. We now apply the definition of $\left(\frac{c}{c+d}\right)_*$ and various parts of Lemma 1 to obtain

$$\left(\frac{c}{c+d}\right)_* = \left(\frac{2^\alpha}{|c+d|}\right)\left(\frac{c_1}{|c+d|}\right)(-1)^{\frac{\operatorname{sign} c-1}{2}\frac{\operatorname{sign}(d+c)-1}{2}}$$

$$= \left(\frac{2}{|c+d|}\right)^\alpha\left(\frac{c+d}{|c_1|}\right)(-1)^{\frac{c_1-1}{2}\frac{c+d-1}{2}} = \left(\frac{2}{|c+d|}\right)^\alpha\left(\frac{d}{|c_1|}\right)(-1)^{\frac{c_1-1}{2}\frac{c+d-1}{2}}$$

Here we have used Lemma 1(g) in the form

$$\left(\frac{n}{|m|}\right)\left(\frac{m}{|n|}\right) = (-1)^{\frac{\operatorname{sign} n - 1}{2}\frac{\operatorname{sign} m - 1}{2}}(-1)^{\frac{n-1}{2}\frac{m-1}{2}}.$$

Proceeding further in this same vein we obtain

$$\left(\frac{c}{c+d}\right)_* = \left(\frac{2}{|c+d|}\right)^{\alpha}\left(\frac{c_1}{|d|}\right)(-1)^{\frac{\operatorname{sign} c_1 - 1}{2}\frac{\operatorname{sign} d - 1}{2}}(-1)^{\frac{c_1-1}{2}\left(\frac{c}{2}+d-1\right)}$$

$$= \left(\frac{2}{|c+d|}\right)^{\alpha}\left(\frac{c_1}{|d|}\right)_*(-1)^{\frac{c_1-1}{2}\frac{c}{2}},$$

since d is odd. But

$$\left(\frac{2}{|c+d|}\right) = (-1)^{\frac{d^2-1}{8}}(-1)^{\frac{c^2+2cd}{8}} = \left(\frac{2}{|d|}\right)(-1)^{\frac{c^2+2cd}{8}},$$

so that

$$\left(\frac{c}{c+d}\right)_* = \left(\frac{2}{|d|}\right)^{\alpha}\left(\frac{c_1}{d}\right)_*(-1)^{\frac{c_1-1}{2}\frac{c}{2}}(-1)^{\frac{c^2+2cd}{8}\alpha}$$

$$= \left(\frac{c}{d}\right)_*(-1)^{\frac{c_1-1}{2}\frac{c}{2}}(-1)^{\frac{c^2+2cd}{8}\alpha}.$$

Thus the proposed formula for $v_\eta(M')$ becomes, after suitable rearrangements,

$$\left(\frac{c}{d}\right)_* \exp\left\{\frac{\pi i}{12}[(a+d)c - bd(c^2-1) + 3d - 2 - 3cd]\right\}\exp\left(\frac{\pi i}{12}E\right)$$

$$= v_\eta(M')\exp\left(\frac{\pi i}{12}E\right),$$

where $E = -3c^2 - (ac+2bc)(c^2-1) + 3c + 3c(c_1-1) + \frac{3}{2}c^2\alpha + 3dc\alpha$. Thus to verify the formula for $v_\eta(M')$ it is sufficient to prove that $E \equiv 0 \pmod{24}$. It is very easy to see that $3|E$. Thus it is sufficient to prove that $8|E$. The proof of this latter fact is split into the three cases $\alpha \geqslant 3$, $\alpha = 2$, and $\alpha = 1$. If $\alpha \geqslant 3$, then $8|c$ and it is immediate that $8|E$. If $\alpha = 2$, then $E = c\left\{\frac{3c}{4} + 6d - (a+2b)(c^2-1)\right\} \equiv 0 \pmod 8$, since $c/4$ and a are both odd. If $\alpha = 1$, then $E = c\{3d - (a+2b)(c^2-1)\}$. If b is even, then $a \equiv d \equiv \pm 1 \pmod 4$. If b is odd, then $a \equiv -d \pm 1 \pmod 4$. In either case $E/c \equiv 3d + a + 2b \equiv 0 \pmod 4$. Once again $8|E$ and the proof of case 1 is complete.

Case 2. $M' = MS^{-1} = \begin{pmatrix} a & b-a \\ c & d-c \end{pmatrix}$. As in case 1 the consistency condition (1) implies that $v_\eta(M') = v_\eta(M)v_\eta(S^{-1}) = v_\eta(M)e^{-\pi i/12}$.

(a) Suppose c is odd. The inductive hypothesis gives

$$v_\eta(M') = \left(\frac{d}{c}\right)^* \exp\left\{\frac{\pi i}{12}[(a+d)c - bd(c^2-1) - 3c]\right\}e^{-\pi i/12}$$

$$= \left(\frac{d}{c}\right)^* \exp\left\{\frac{\pi i}{12}[(a+d)c - bd(c^2-1) - 3c - 1]\right\}.$$

We now compare this with the proposed formula for $v_\eta(M')$, which is

$$\left(\frac{d-c}{c}\right)^* \exp\left\{\frac{\pi i}{12}[(a+d-c)c - (b-a)(d-c)(c^2-1) - 3c]\right\}$$

$$= \left(\frac{d}{c}\right)^* \exp\left\{\frac{\pi i}{12}[(a+d)c - bd(c^2-1) - 3c - 1]\right\} \exp\left(\frac{\pi i}{12}E\right),$$

where $E = 1 - c^2 - (ac - ad - bc)(c^2-1)$. It is sufficient to prove that $E \equiv 0 \pmod{24}$. But using $ad = 1 + bc$, we get $E = -(ac - 2bc)(c^2-1) \equiv 0 \pmod{24}$.

(b) Suppose c is even. By the induction hypothesis

$$v_\eta(M') = \left(\frac{c}{d}\right)_* \exp\left\{\frac{\pi i}{12}[(a+d)c - bd(c^2-1) + 3d - 4 - 3cd]\right\}.$$

On the other hand, the proposed formula for $v_\eta(M')$ is

$$\left(\frac{c}{d-c}\right)_* \exp\left\{\frac{\pi i}{12}[(a+d-c)c - (b-a)(d-c)(c^2-1) + 3(d-c) - 3 - 3c(d-c)]\right\}.$$

Now if $c = 0$, then $a = d = \pm 1$ and

$$v_\eta(M') = \left(\frac{0}{\pm 1}\right)_* \exp\left\{\frac{\pi i}{12}(bd + 3d - 4)\right\},$$

while, on the other hand, the proposed formula becomes

$$\left(\frac{0}{\pm 1}\right)_* \exp\left\{\frac{\pi i}{12}[(b-a)d + 3d - 3]\right\} = \left(\frac{0}{\pm 1}\right)_* \exp\left\{\frac{\pi i}{12}(bd + 3d - 4)\right\}.$$

The formula is therefore correct if $c = 0$. Suppose then that $c \neq 0$ and write $c = 2^\alpha c_1$, where c_1 is odd and $\alpha \geqslant 1$. A calculation using Lemma 1 and similar to that in case 1(b) yields

$$\left(\frac{c}{d-c}\right)_* = \left(\frac{c}{d}\right)_* (-1)^{\frac{c_1-1}{2}\left(-\frac{c}{2}\right)}(-1)^{\frac{c^2-2cd}{8}\alpha}.$$

The proposed formula thus becomes

$$\left(\frac{c}{d}\right)_* \exp\left\{\frac{\pi i}{12}[(a+d)c - bd(c^2-1) + 3d - 4 - 3cd]\right\} \exp\left(\frac{\pi i}{12}E\right)$$

$$= v_\eta(M') \exp\left(\frac{\pi i}{12}E\right),$$

where $E = 3c^2 - (ac - 2bc)(c^2-1) - 3c - 3c(c_1 - 1) + \frac{3}{2}c^2\alpha - 3cd\alpha$. But, as in case 1(b), a calculation shows that $E \equiv 0 \pmod{24}$ and this completes the proof of case 2.

Case 3. $M' = MT = \begin{pmatrix} a & b \\ c & d \end{pmatrix}\begin{pmatrix} 0 & -1 \\ 1 & 0 \end{pmatrix} = \begin{pmatrix} b & -a \\ d & -c \end{pmatrix}$. In this case the consistency condition (1) implies that

$$v_\eta(M')(d\tau - c)^{1/2} = v_\eta(M)v_\eta(T)(-c/\tau + d)^{1/2}\tau^{1/2}.$$

We also need the fact that

$$(2) \qquad \frac{(-c/\tau + d)^{1/2}\tau^{1/2}}{(d\tau - c)^{1/2}} = (-1)^{\frac{-\operatorname{sign} c - 1}{2}\frac{\operatorname{sign} d - 1}{2}} \qquad \text{if } c \neq 0, d \neq 0.$$

It is clear that the left-hand side of (2) has absolute value 1. The proof of (2) can then be carried out by simply calculating the argument of the left-hand side. To do this recall that, by our branch convention,

$$0 < \arg\tau < \pi, \qquad 0 < |\arg(d\tau - c)| < \pi \qquad \text{and} \qquad 0 < |\arg(-c/\tau + d)| < \pi.$$

A consideration of four cases arranged according to the sign of c and the sign of d now easily completes the proof of (2). From (2) it follows that

$$(3) \qquad v_\eta(M') = v_\eta(M)e^{-\pi i/4}(-1)^{\frac{-\operatorname{sign} c - 1}{2}\frac{\operatorname{sign} d - 1}{2}} \qquad \text{if } c \neq 0, d \neq 0.$$

(a) Suppose c and d are both odd. Then in particular $c \neq 0$, $d \neq 0$ and (3) applies. By the induction hypothesis

$$v_\eta(M) = \left(\frac{d}{c}\right)^* \exp\left\{\frac{\pi i}{12}[(a+d)c - bd(c^2-1) - 3c]\right\},$$

so that by (3),

$$v_\eta(M') = \left(\frac{d}{c}\right)^* \exp\left\{\frac{\pi i}{12}[(a+d)c - bd(c^2-1) - 3c]\right\} e^{-\pi i/4}(-1)^{\frac{-\operatorname{sign} c - 1}{2}\frac{\operatorname{sign} d - 1}{2}}.$$

The proposed formula for $v_\eta(M')$, on the other hand, gives

$$\left(\frac{-c}{d}\right)^* \exp\left\{\frac{\pi i}{12}[(b-c)d - ac(d^2-1) - 3d]\right\},$$

since d is odd. By Lemma 1(g),

$$\left(\frac{-c}{d}\right)^* = \left(\frac{d}{c}\right)^*(-1)^{\frac{-\operatorname{sign} c-1}{2}\frac{\operatorname{sign} d-1}{2}}(-1)^{\frac{-c-1}{2}\frac{d-1}{2}}.$$

Thus the proposed formula becomes $v_\eta(M')\exp\left(\frac{\pi i}{12}E\right)$, where

$$\begin{aligned} E &= bd - cd - acd^2 + ac - 3d - ac - dc + bdc^2 - bd + 3c + 3 \\ &\quad -3(c+1)(d-1) = -6(c+1)(d-1) \equiv 0 \pmod{24}. \end{aligned}$$

It follows that $v_\eta(M')$ agrees with the proposed formula.

(b) Suppose c is odd, but d is even. Then $c \neq 0$. If $d = 0$, then $b = -c = \pm 1$. The consistency condition (1) implies that

$$v_\eta(M') = v_\eta(M)v_\eta(T)\frac{(-c/\tau)^{1/2}\tau^{1/2}}{(-c)^{1/2}}.$$

If we consider the two cases $c = 1$ and $c = -1$ and use our branch convention, we find that

$$\frac{(-c/\tau)^{1/2}\tau^{1/2}}{(-c)^{1/2}} = (-1)^{\frac{c+1}{2}} = -c.$$

Thus

$$v_\eta(M') = v_\eta(M)e^{-\pi i/4}e^{(\pi i/2)(c+1)}.$$

By the induction hypothesis

$$v_\eta(M) = \left(\frac{0}{c}\right)^* \exp\left\{\frac{\pi i}{12}(ac - 3c)\right\},$$

so that

$$v_\eta(M') = \exp\left\{\frac{\pi i}{12}(ac + 3c + 3)\right\}.$$

The proposed formula for $v_\eta(M')$ is

$$\left(\frac{d}{-c}\right)_* \exp\left\{\frac{\pi i}{12}[(b-c)0 - ac(-1) + 3(-c) - 3]\right\}$$

$$= \left(\frac{0}{-c}\right)_* \exp\left\{\frac{\pi i}{12}(ac - 3c - 3)\right\} = (-1)^{\frac{c+1}{2}} \exp\left\{\frac{\pi i}{12}(ac - 3c - 3)\right\}$$

$$= \exp\left\{\frac{\pi i}{12}(ac + 3c + 3)\right\} = v_\eta(M').$$

We may now assume that $d \neq 0$. In this case (3) applies and we have

$$v_\eta(M') = v_\eta(M)e^{-\pi i/4}(-1)^{\frac{-\operatorname{sign} c - 1}{2}\frac{\operatorname{sign} d - 1}{2}}.$$

The induction hypothesis applied to $v_\eta(M)$ gives

$$v_\eta(M') = \left(\frac{d}{c}\right)^* \exp\left\{\frac{\pi i}{12}[(a+d)c - bd(c^2 - 1) - 3c]\right\} e^{-\pi i/4}(-1)^{\frac{-\operatorname{sign} c - 1}{2}\frac{\operatorname{sign} d - 1}{2}}.$$

The proposed formula for $v_\eta(M')$ is

$$\left(\frac{d}{-c}\right)^* \exp\left\{\frac{\pi i}{12}[(b-c)d - ac(d^2 - 1) - 3c - 3 + 3cd]\right\}$$

$$= \left(\frac{d}{c}\right)^*(-1)^{\frac{-\operatorname{sign} c - 1}{2}\frac{\operatorname{sign} d - 1}{2}} \exp\left\{\frac{\pi i}{12}[(b-c)d - ac(d^2 - 1)\right.$$

$$\left. -3c - 3 + 3cd]\right\}.$$

This is the same as

$$v_\eta(M')e^{\pi i/4} \exp\left\{\frac{-\pi i}{12}(2cd - 3cd + 3 + acd^2 - bdc^2)\right\}$$

$$= v_\eta(M')e^{\pi i/4}e^{-\pi i 3/12} = v_\eta(M'),$$

and the proof is complete for Case 3(b).

(c) Suppose c is even. Then d is odd and in particular $d \neq 0$. If $c = 0$, then $d = \pm 1$ and (1) becomes

$$v_\eta(M')(d\tau)^{1/2} = v_\eta(M)v_\eta(T)d^{1/2}\tau^{1/2}.$$

But $(d\tau)^{1/2} = d^{1/2}\tau^{1/2}$, by our branch convention, so that, in this case,

$$v_\eta(M') = v_\eta(M)v_\eta(T) = v_\eta(M)e^{-\pi i/4}.$$

By the induction hypothesis

$$v_\eta(M') = \left(\frac{0}{d}\right)_* \exp\left\{\frac{\pi i}{12}(bd + 3d - 3)\right\} e^{-\pi i/4} = \left(\frac{0}{d}\right)_* \exp\left\{\frac{\pi i}{12}(bd + 3d - 6)\right\}$$

$$= (-1)^{\frac{d-1}{2}} \exp\left\{\frac{\pi i}{12}(bd + 3d - 6)\right\} = \exp\left\{\frac{\pi i}{12}(bd + 9d - 12)\right\}.$$

The proposed formula for $v_\eta(M')$ is

$$\left(\frac{0}{d}\right)^* \exp\left\{\frac{\pi i}{12}(bd - 3d)\right\} = \exp\left\{\frac{\pi i}{12}(bd - 3d)\right\}.$$

Clearly $bd + 9d - 12 \equiv bd - 3d \pmod{24}$. Thus $v_\eta(M')$ agrees with the proposed formula.

Assume now that $c \neq 0$. Then (3) applies so that, by induction,

$$v_\eta(M') = v_\eta(M)v_\eta(T)(-1)^{\frac{-\operatorname{sign} c - 1}{2}\frac{\operatorname{sign} d - 1}{2}}$$

$$= \left(\frac{c}{d}\right)_* \exp\left\{\frac{\pi i}{12}[(a + d)c - bd(c^2 - 1) + 3d - 3 - 3cd]\right\} e^{-\pi i/4}$$

$$\times (-1)^{\frac{-\operatorname{sign} c - 1}{2}\frac{\operatorname{sign} d - 1}{2}}.$$

But

$$\left(\frac{c}{d}\right)_* (-1)^{\frac{-\operatorname{sign} c - 1}{2}\frac{\operatorname{sign} d - 1}{2}} = \left(\frac{c}{d}\right)^* (-1)^{\frac{\operatorname{sign} d - 1}{2}},$$

so that

$$v_\eta(M') = \left(\frac{c}{d}\right)^* \exp\left\{\frac{\pi i}{12}[(a + d)c - bd(c^2 - 1) + 3d - 3 - 3cd]\right\} e^{-\pi i/4}(-1)^{\frac{\operatorname{sign} d - 1}{2}}.$$

The proposed formula for $v_\eta(M')$ is

$$\left(\frac{-c}{d}\right)^* \exp\left\{\frac{\pi i}{12}[(b - c)d - ac(d^2 - 1) - 3d]\right\}$$

$$= (-1)^{\frac{|d| - 1}{2}} \left(\frac{c}{d}\right)^* \exp\left\{\frac{\pi i}{12}[(b - c)d - ac(d^2 - 1) - 3d]\right\},$$

by Lemma 1(d). Thus the proposed formula equals

$$v_\eta(M')(-1)^{\frac{|d|-1}{2}}(-1)^{\frac{\operatorname{sign} d-1}{2}}\exp\left\{\frac{-\pi i}{12}(2cd+6d-3-3cd+acd^2-bdc^2)\right\}e^{\pi i/4}$$
$$=v_\eta(M')(-1)^{\frac{|d|-1}{2}}(-1)^{\frac{\operatorname{sign} d-1}{2}}\exp\left\{\frac{\pi i}{12}(3d-6)\right\}e^{\pi i/4}$$
$$=v_\eta(M')(-1)^{\frac{|d|-1}{2}}(-1)^{\frac{\operatorname{sign} d-1}{2}}e^{(\pi i/2)(1-d)}=v_\eta(M').$$

This completes the proof of Case 3(c) and, with it, of Theorem 2.

3. PROOF OF THEOREM 3

In the proof of Theorem 13 of Chapter 3 [formula (17)] we observed that v_ϑ is given in terms of v_η in the following way. If $M=\begin{pmatrix}a & b\\ c & d\end{pmatrix}\in\Gamma_\vartheta$, then $v_\vartheta(M)=v_\eta^2(M_1)v_\eta(M_2)^{-1}e^{\pi i/12}$, where

$$M_1=\begin{pmatrix}a+c & \dfrac{b-c}{2}+\dfrac{d-a}{2}\\ 2c & d-c\end{pmatrix},\qquad M_2=\begin{pmatrix}a+c & b+d\\ c & d\end{pmatrix}.$$

There are two cases to consider since either $a\equiv d\equiv 0$, $b\equiv c\equiv 1 \pmod 2$ or $a\equiv d\equiv 1$, $b\equiv c\equiv 0 \pmod 2$.

Case 1. $a\equiv d\equiv 0$, $b\equiv c\equiv 1 \pmod 2$. By Theorem 2 we have

$$v_\eta^2(M_1)=\left(\frac{2c}{d-c}\right)_*^2\exp\left\{\frac{\pi i}{6}\left[(a+d)2c-\frac{(d-c)}{2}(b-c+d-a)(4c^2-1)\right.\right.$$
$$\left.\left.+3d-3c-3-6c(d-c)\right]\right\}$$
(4)
$$=\exp\left\{\frac{\pi i}{6}\left[(a+d)2c-\frac{(d-c)}{2}(b-c+d-a)(4c^2-1)\right.\right.$$
$$\left.\left.+3d-3c-3-6c(d-c)\right]\right\},$$

and

$$v_\eta(M_2)=\left(\frac{d}{c}\right)^*\exp\left\{\frac{\pi i}{12}[(a+c+d)c-(b+d)d(c^2-1)-3c]\right\}.$$

Thus

$$v_\vartheta(M)=v_\eta^2(M_1)v_\eta(M_2)^{-1}e^{\pi i/12}=\left(\frac{d}{c}\right)^*e^{(\pi i/12)E},$$

where

$$E = 4ac + cd + 6d - 3c - 6 + 12c^2 - ad - bc - 3bdc^2$$
$$+ 4dc^3 - 4c^2d^2 + 4adc^2 + 4bc^3 - 4c^4 + 4c^3d - 4ac^3 + d^2c^2 + 1.$$

Calculating modulo 24, we obtain

$$\begin{aligned} E &\equiv 4ac + cd + 6d - 3c + 6 - 2bc - 3bdc^2 + 8dc^3 - 4c^2d^2 + 4adc^2 \\ &\quad +4bc^3 - 4c^4 - 4ac^3 + d^2c^2 \\ &\equiv 4ac(1 - c^2) + 9cd + 8d(c^2 - 1) - 3c^2d^2 + 4c^2(bc + 1) - 4c^4 \\ &\quad +4bc^3 - 3bdc^2 + 6d - 3c + 6 - 2bc \\ &\equiv 9cd - 3c^2d^2 + 4c^2(1 - c^2) + 8bc^3 - 3bdc^2 + 6d - 3c + 6 - 2bc \\ &\equiv 9cd - 3c^2d^2 + 8bc(c^2 - 1) + 6bc - 3bdc^2 + 6d - 3c + 6 \\ &\equiv 9cd - 3c^2d^2 + 6(bc + 1) - 3bdc^2 + 6d - 3c \\ &\equiv 6d(c + 1) + 3cd - 3c^2d^2 - 3bdc^2 - 3c \\ &\equiv 3cd(1 - cd - bc) - 3c \equiv -3c \pmod{24}, \end{aligned}$$

since $ad - bc = 1$ and $a \equiv d \equiv 0$, $b \equiv c \equiv 1 \pmod 2$. It follows that

$$v_\vartheta(M) = \left(\frac{d}{c}\right)^* e^{-\pi i c/4}.$$

Case 2. $a \equiv d \equiv 1$, $b \equiv c \equiv 0 \pmod 2$. As in Case 1, the expression (4) holds for $v_\eta(M_1)^2$. In this case, however, we have

$$v_\eta(M_2) = \left(\frac{c}{d}\right)_* \exp\left\{\frac{\pi i}{12}[(a + c + d)c - (b + d)d(c^2 - 1) + 3d - 3 - 3cd]\right\}.$$

Then we have

$$v_\vartheta(M) = v_\eta^2(M_1)v_\eta(M_2)^{-1}e^{\pi i/12} = \left(\frac{c}{d}\right)_* e^{(\pi i/12)E},$$

where

$$E = 4ac - 8cd + 3d - 6c - 2 + 12c^2 - ad - bc - 3bdc^2$$
$$+4dc^3 + 4c^3d - 3c^2d^2 + 4adc^2 + 4bc^3 - 4c^4 - 4dc^3 - 4ac^3.$$

Again a calculation modulo 24 shows that

$$\begin{aligned} E &\equiv 4ac + 4cd + 3d - 6c - 3 - 2bc - 3bdc^3 - 3d^2c^2 + 8c^3d + 4adc^2 \\ &\quad + 4bc^3 - 4c^4 - 4ac^3 \\ &\equiv 4ac(1 - c^2) + 8cd(c^2 - 1) + 12cd + 4bc(c^2 - 1) + 2bc + 3d - 6c \\ &\quad - 3 - 3bdc^3 - 3d^2c^2 + 4adc^2 - 4c^4 \\ &\equiv 2bc + 3d - 6c - 3 - 3d^2c^2 + 4c^2 + 4bc^3 - 4c^4 \\ &\equiv 4c^2(1 - c^2) + 4bc(c^2 - 1) + 6bc + 3d - 6c - 3 - 3d^2c^2 \\ &\equiv 3d - 6c - 3 - 3d^2c^2 \equiv 3d - 3 \pmod{24}. \end{aligned}$$

It follows that

$$v_\vartheta(M) = \left(\frac{c}{d}\right)_* e^{\pi i(d-1)/4},$$

and the proof is complete.

Chapter 5

SUMS OF SQUARES

1. STATEMENT OF RESULTS

In this chapter we apply the theory of the function $\vartheta(\tau)$ to derive information concerning the number-theoretic function $r_s(m)$ which was defined in Chapter 3, Section 4. Before we can describe the main results we must introduce some definitions and notation.

As before, we denote the multiplier system of $\vartheta(\tau)$ by v_ϑ. Let s be an integer and put $v_\vartheta^s = v_s$. Since v_ϑ is a multiplier system for the group Γ_ϑ and the degree $-\frac{1}{2}$, it follows easily that v_s is a multiplier system for the group Γ_ϑ and the degree $-s/2$. That is (see Chapter 2, Section 1),

(1) $$v_s(M_1M_2)(c_3\tau + d_3)^{s/2} = v_s(M_1)v_s(M_2)(c_1M_2\tau + d_1)^{s/2}(c_2\tau + d_2)^{s/2},$$

for all M_1 and M_2 in Γ_ϑ, where

$$M_1 = \begin{pmatrix} * & * \\ c_1 & d_1 \end{pmatrix}, \qquad M_2 = \begin{pmatrix} * & * \\ c_2 & d_2 \end{pmatrix}, \qquad M_1M_2 = \begin{pmatrix} * & * \\ c_3 & d_3 \end{pmatrix}.$$

Recall also that $v_\vartheta(S^2) = 1$; thus $v_s(S^2) = 1$. It then follows from (1) that

(2) $$v_s(MS^{2q}) = v_s(S^{2q}M) = v_s(M),$$

for any integer q and any $M \in \Gamma_\vartheta$. Furthermore, since v_s is the multiplier system connected with the nontrivial function $\vartheta(\tau)$, we also have

(3) $$v_s(-M)(-c\tau - d)^{s/2} = v_s(M)(c\tau + d)^{s/2},$$

for every $M = \begin{pmatrix} * & * \\ c & d \end{pmatrix} \in \Gamma_\vartheta$.

Definition. (a) If c and m are positive integers, let

$$B_c(m) = c^{-s/2} \sum_{0 \leq h < 2c}^{*} \bar{v}_s(M_{c,h}) \exp\left(\frac{\pi i m h}{c}\right), \tag{4}$$

where $M_{c,h} = \begin{pmatrix} * & * \\ c & h \end{pmatrix}$ is *any* element of Γ_ϑ with lower row c, h, and Σ^* indicates that we are to sum only over integers h such that $(h,c) = 1$ and h is of parity opposite to the parity of c.

(b) If m is a positive integer, put

$$\rho_s(m) = \frac{e^{-\pi i s/4}\pi^{s/2}}{\Gamma(s/2)} m^{s/2-1} \sum_{c=1}^{\infty} B_c(m). \tag{5}$$

This is sometimes referred to as *the singular series*.

Remarks. 1. $B_c(m)$, of course, depends upon s as well as c and m.

2. There are two slight difficulties that arise in connection with our definition of $B_c(m)$. First, we must show that given any pair of integers h, c of opposite parity such that $(h,c) = 1$, there exists an $M_{c,h} \in \Gamma_\vartheta$. The second problem is that the element $M_{c,h}$ is *not unique.* Nevertheless, we can still show that $B_c(m)$ is uniquely defined. Both of these difficulties are resolved in

Lemma 1. (a) Given integers h and c of opposite parity such that $(h,c) = 1$, then there exists integers a and b such that $\begin{pmatrix} a & b \\ c & h \end{pmatrix} \in \Gamma_\vartheta$.

(b) If $M = \begin{pmatrix} a & b \\ c & h \end{pmatrix} \in \Gamma_\vartheta$ and $M' = \begin{pmatrix} a' & b' \\ c & h \end{pmatrix} \in \Gamma_\vartheta$, then $v_s(M) = v_s(M')$.

Proof. (a) Since $(c,h) = 1$, there exist integers a, b such that $ah - bc = 1$. Suppose c is odd and h is even. Since $ah - bc = 1$, it follows that b is odd. On the other hand, a may be even or odd. If a is even, then $\begin{pmatrix} a & b \\ c & h \end{pmatrix} \in \Gamma_\vartheta$ (see Corollary 4 of Chapter 1), and we are done. If a is odd, we replace a by $a + c$ and b by $b + h$; then $(a + c)h - c(b + h) = 1$, $a + c$ is even, and $b + h$ is odd. Thus $\begin{pmatrix} a + c & b + h \\ c & h \end{pmatrix} \in \Gamma_\vartheta$. If c is even and h odd, we proceed similarly.

(b) $MM'^{-1} = \begin{pmatrix} a & b \\ c & h \end{pmatrix}\begin{pmatrix} h & -b' \\ -c & a' \end{pmatrix} = \begin{pmatrix} 1 & a'b - ab' \\ 0 & 1 \end{pmatrix}$. Also $MM'^{-1} \in \Gamma_\vartheta$ so that $a'b - ab'$ is even. Hence $MM'^{-1} = S^{2q}$ for some integer q. Thus by (2),

$$v_s(M) = v_s(S^{2q}M') = v_s(M').$$

This completes the proof.

3. The definition of $\rho_s(m)$ makes sense only if the infinite series on the right-hand side of (5) converges. Since, obviously, $|B_c(m)| \leqslant 2c^{1-s/2}$, it follows that the series converges absolutely for $s > 4$. Here we deal only with integers $s \geqslant 5$.

The main results of this chapter are contained in the following two theorems.

Theorem 2. If s is an integer $\geqslant 5$, we have

$$r_s(m) = \rho_s(m) + O(m^{s/4}), \qquad \text{as } m \to +\infty.$$

Theorem 3. If s is an integer such that $5 \leqslant s \leqslant 8$, then $r_s(m) = \rho_s(m)$ for all $m \geqslant 1$.

Theorems 2 and 3 are remarkable in that they express the purely number-theoretic function $r_s(m)$ in terms of the complicated analytic expression $\rho_s(m)$. The remainder of this chapter will be devoted to the proofs of Theorems 2 and 3 and related results. It should be pointed out that Theorem 3 is true also when $s = 3$ and $s = 4$, but the proof in these cases (especially in the case $s = 3$) is much more difficult and we do not give it. For the proof in these cases the reader is referred to the thesis of P. T. Bateman [*Trans. Am. Math. Soc.*, *71* (1951), pp. 70–101].

2. LIPSCHITZ SUMMATION FORMULA

As a first step toward proving Theorems 2 and 3 we state and prove

Theorem 4 (Lipschitz Summation Formula). For $\tau \in \mathscr{H}$ and $\lambda > 1$, we have

$$\sum_{m=-\infty}^{\infty} (2m + \tau)^{-\lambda} = \frac{e^{-\pi i\lambda/2}\pi^{\lambda}}{\Gamma(\lambda)} \sum_{n=1}^{\infty} n^{\lambda-1} e^{\pi i\tau n}.$$

In the proof of Theorem 4 we need two preliminary lemmas.

Lemma 5. Let N and ρ be positive real numbers and let $D_{N,\rho}$ denote the following contour traversed in the counterclockwise direction (Figure 3):

the negative real axis from $-N$ to $-\rho$ (argument $= -\pi$);
the circle of radius ρ centered at 0;
the negative real axis from $-\rho$ to $-N$ (argument $= \pi$).

Then with fixed ρ, n a positive integer, and arbitrary complex λ we have

$$\lim_{N\to+\infty} \left\{ -i \int_{D_{N,\rho}} z^{-\lambda} e^{2\pi n z}\, dz \right\} = \frac{(2\pi)^{\lambda} n^{\lambda-1}}{\Gamma(\lambda)}.$$

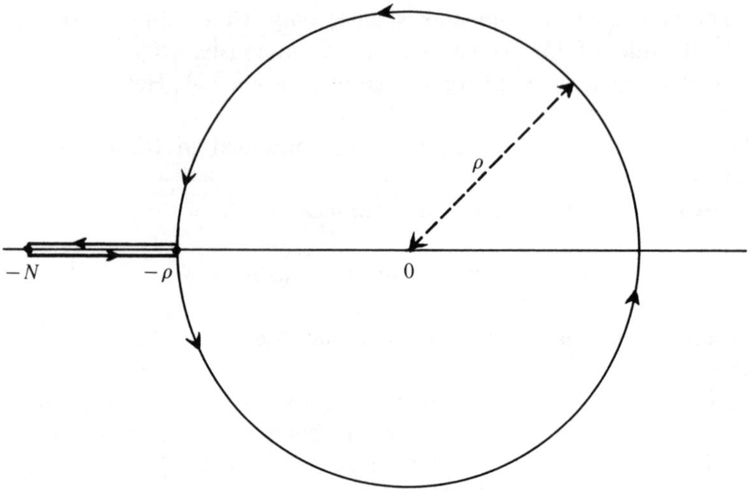

Figure 3. The contour $D_{N,\rho}$.

Proof. We have

$$\int_{D_{N,\rho}} t^{\lambda-1}e^{t}\,dt = (e^{\pi i(\lambda-1)} - e^{-i\pi(\lambda-1)})\int_{-\rho}^{-N}(-t)^{\lambda-1}e^{t}\,dt$$
$$+ i\rho^{\lambda}\int_{-\pi}^{\pi}\exp(i\lambda\theta + \rho e^{i\theta})\,d\theta$$
$$= 2i\sin\pi\lambda\int_{\rho}^{N} t^{\lambda-1}e^{-t}\,dt + i\rho^{\lambda}\int_{-\pi}^{\pi}\exp(i\lambda\theta + \rho e^{i\theta})\,d\theta.$$

Now assume that $\operatorname{Re}\lambda > 0$ and let $\rho \to 0+$. Since $|\exp(i\lambda\theta + \rho e^{i\theta})| = \exp(\rho\cos\theta - \theta\operatorname{Im}\lambda)$, the second term above approaches 0. Thus for $\operatorname{Re}\lambda > 0$,

$$\lim_{\rho\to 0+}\int_{D_{N,\rho}} t^{\lambda-1}e^{t}\,dt = 2i\sin\pi\lambda\int_{0}^{N} t^{\lambda-1}e^{-t}\,dt.$$

However, by Cauchy's Theorem, $\int_{D_{N,\rho}} t^{\lambda-1}e^{t}\,dt$ is clearly *independent* of ρ, as long as $\rho > 0$. Therefore, for fixed $\rho > 0$,

$$\int_{D_{N,\rho}} t^{\lambda-1}e^{t}\,dt = 2i\sin\pi\lambda\int_{0}^{N} t^{\lambda-1}e^{-t}\,dt,$$

provided $\operatorname{Re}\lambda > 0$. It follows that for $\operatorname{Re}\lambda > 0$,

$$\lim_{N\to+\infty}\int_{D_{N,\rho}} t^{\lambda-1}e^t\,dt = 2i\sin\pi\lambda\int_0^\infty t^{\lambda-1}e^{-t}\,dt = 2i\sin\pi\lambda\,\Gamma(\lambda).$$

Using the well-known formula

$$\sin\pi\lambda = \pi/\Gamma(\lambda)\Gamma(1-\lambda),$$

we obtain

$$\lim_{N\to+\infty}\int_{D_{N,\rho}} t^{\lambda-1}e^t\,dt = 2\pi i/\Gamma(1-\lambda) \tag{6}$$

as long as $\operatorname{Re}\lambda > 0$.

We wish to obtain (6) for all complex λ. One possibility is to observe that $1/\Gamma(z)$ is an entire function of z so that the right-hand side of (6) is an entire function of λ, while, on the other hand, the left-hand side of (6) is an entire function of λ by the Lebesgue dominated convergence theorem. It then follows that (6) holds in the entire λ-plane by the identity theorem for analytic functions.

A more elementary, though slightly longer, proof can be carried out by induction on $\operatorname{Re}\lambda$. The details follow. Suppose a is a nonnegative integer and suppose (6) holds for $\operatorname{Re}\lambda > -a$. Let λ be such that $-a-1 < \operatorname{Re}\lambda \leqslant -a$, but $\lambda \neq -a$. Consider $\int_{D_{N,\rho}} t^{\lambda-1}e^t\,dt$ and integrate by parts to obtain

$$\int_{D_{N,\rho}} t^{\lambda-1}e^t\,dt = \lambda^{-1}e^t t^\lambda]_{D_{N,\rho}} - \frac{1}{\lambda}\int_{D_{N,\rho}} t^\lambda e^t\,dt$$

$$= \lambda^{-1}N^\lambda e^{-N}(e^{i\pi\lambda} - e^{-i\pi\lambda}) - \frac{1}{\lambda}\int_{D_{N,\rho}} t^{(\lambda+1)-1}e^t\,dt.$$

Since $\operatorname{Re}(\lambda+1) > -a$, the induction hypothesis implies that

$$\lim_{N\to+\infty}\int_{D_{N,\rho}} t^{\lambda-1}e^t\,dt = -\frac{1}{\lambda}\frac{2\pi i}{\Gamma(-\lambda)} = \frac{2\pi i}{\Gamma(1-\lambda)}.$$

Thus, by induction, (6) holds for all complex λ except for $\lambda = 0, -1, -2, \ldots$. For these λ, however, (6) reduces to

$$\int_{C_\rho} t^{\lambda-1}e^t\,dt = \frac{2\pi i}{\Gamma(1-\lambda)} = \frac{2\pi i}{(-\lambda)!}, \tag{7}$$

where C_ρ is the circle of radius ρ centered at 0 and traversed in the counterclockwise direction. Formula (7) follows immediately if we apply the Cauchy integral formula to the function e^t. Thus (6) holds for all complex λ.

In (6) replace λ by $1 - \lambda$ and we obtain

$$\lim_{N\to+\infty}\int_{D_{N,\rho}} t^{-\lambda}e^t\,dt = \frac{2\pi i}{\Gamma(\lambda)},$$

for all complex λ. Making the substitution $t = 2\pi nz$, with n a positive integer, we get

$$\int_{D_{2\pi nN,2\pi n\rho}} t^{-\lambda}e^t\,dt = \int_{D_{N,\rho}} (2\pi nz)^{-\lambda}e^{2\pi nz}(2\pi n)\,dz$$

$$= (2\pi n)^{1-\lambda}\int_{D_{N,\rho}} z^{-\lambda}e^{2\pi nz}\,dz.$$

Hence

$$\lim_{N\to+\infty}\int_{D_{N,\rho}} z^{-\lambda}e^{2\pi nz}\,dz = (2\pi n)^{\lambda-1}\lim_{N\to+\infty}\int_{D_{2\pi nN,2\pi n\rho}} t^{-\lambda}e^t\,dt$$

$$= \frac{i(2\pi)^{\lambda}n^{\lambda-1}}{\Gamma(\lambda)},$$

and the proof is complete.

Remark. The reader unfamiliar with the properties of $\Gamma(z)$ is referred to E. T. Whittaker and G. N. Watson, *A Course of Modern Analysis* (4th ed.; New York: Cambridge University Press, 1927), pp. 235–264.

Lemma 6. (a) Let n be a negative integer and let λ be a complex number with $\operatorname{Re}\lambda > 0$. Then for any $t > 0$,

$$\int_{t-i\infty}^{t+i\infty} z^{-\lambda}e^{2\pi nz}\,dz = 0.$$

(b) If $\operatorname{Re}\lambda > 1$ and $t > 0$, then

$$\int_{t-i\infty}^{t+i\infty} z^{-\lambda}\,dz = 0.$$

Proof. (a) For the moment we take it for granted that the improper integral $\int_{t-i\infty}^{t+i\infty} z^{-\lambda}e^{2\pi nz}\,dz$ converges, for any integer n. A proof of this fact will emerge in due course. We first claim that for $t_1,t_2 > 0$,

$$\int_{t_1-i\infty}^{t_1+i\infty} z^{-\lambda}e^{2\pi nz}\,dz = \int_{t_2-i\infty}^{t_2+i\infty} z^{-\lambda}e^{2\pi nz}\,dz. \tag{8}$$

Without loss of generality we assume that $t_2 > t_1$. Cauchy's theorem implies that $\int_C z^{-\lambda}e^{2\pi nz}\,dz = 0$, where C is the rectangular path with vertices $t_1 + iN, t_2 + iN, t_1 - iM, t_2 - iM$, traversed in the counterclockwise direction. Here M and N are positive real numbers. On the upper horizontal segment of C, we have

$$|z^{-\lambda}e^{2\pi nz}| \leqslant e^{2\pi nt_1}e^{-(\mathrm{Re}\,\lambda)\log N}e^{(\pi/2)|\mathrm{Im}\,\lambda|},$$

so that $z^{-\lambda}e^{2\pi nz} \to 0$ as $N \to +\infty$. On the lower horizontal segment of C,

$$|z^{-\lambda}e^{2\pi nz}| \leqslant e^{2\pi nt_1}e^{-(\mathrm{Re}\,\lambda)\log M}e^{(\pi/2)|\mathrm{Im}\,\lambda|},$$

so that also here $z^{-\lambda}e^{2\pi nz} \to 0$ as $M \to +\infty$. Since the length of each of these horizontal segments is $t_2 - t_1$, it follows that the integral along each approaches 0 as $M,N \to +\infty$, and (8) follows.

Now we consider $\int_{t-iM}^{t+iN} z^{-\lambda}e^{2\pi nz}\,dz$ for arbitrary $t > 0$. An integration by parts yields

$$\begin{aligned}\int_{t-iM}^{t+iN} z^{-\lambda}e^{2\pi nz}\,dz &= \left.\frac{z^{-\lambda}}{2\pi n}e^{2\pi nz}\right]_{t-iM}^{t+iN} + \frac{\lambda}{2\pi n}\int_{t-iM}^{t+iN} z^{-\lambda-1}e^{2\pi nz}\,dz \\ &= \frac{1}{2\pi n}\{(t+iN)^{-\lambda}e^{2\pi n(t+iN)} - (t-iM)^{-\lambda}e^{2\pi n(t-iM)}\} \\ &\quad + \frac{\lambda}{2\pi n}\int_{t-iM}^{t+iN} z^{-\lambda-1}e^{2\pi nz}\,dz.\end{aligned}$$

Since $\mathrm{Re}\,\lambda > 0$ it is a simple matter to verify that the improper integral $\int_{t-i\infty}^{t+i\infty} z^{-\lambda-1}e^{2\pi nz}\,dz$ exists for $t > 0$. Thus for $t > 0$, $\int_{t-i\infty}^{t+i\infty} z^{-\lambda}e^{2\pi nz}\,dz$ exists, and

$$\int_{t-i\infty}^{t+i\infty} z^{-\lambda}e^{2\pi nz}\,dz = \frac{\lambda}{2\pi n}\int_{t-i\infty}^{t+i\infty} z^{-\lambda-1}e^{2\pi nz}\,dz.$$

In particular, then,

$$\text{(9)} \qquad \lim_{t\to+\infty}\int_{t-i\infty}^{t+i\infty} z^{-\lambda}e^{2\pi nz}\,dz = \left(\frac{\lambda}{2\pi n}\right)\lim_{t\to+\infty}\int_{t-i\infty}^{t+i\infty} z^{-\lambda-1}e^{2\pi nz}\,dz.$$

However,

$$\begin{aligned}\left|\int_{t-i\infty}^{t+i\infty} z^{-\lambda-1}e^{2\pi nz}\,dz\right| &\leqslant e^{2\pi nt}\int_{-\infty}^{\infty} |(t+iy)^{-\lambda-1}|\,dy \\ &\leqslant e^{2\pi nt}e^{(\pi/2)|\mathrm{Im}(\lambda+1)|}\left\{\int_{-1}^{1}(\sqrt{t^2+y^2})^{-\mathrm{Re}(\lambda+1)}\,dy\right. \\ &\qquad \left.+2\int_{1}^{\infty}(\sqrt{t^2+y^2})^{-\mathrm{Re}(\lambda+1)}\,dy\right\} \\ &\leqslant e^{2\pi nt}e^{(\pi/2)|\mathrm{Im}(\lambda+1)|}\left\{2t^{-\mathrm{Re}(\lambda+1)} + 2\int_{1}^{\infty} y^{-\mathrm{Re}(\lambda+1)}\,dy\right\} \\ &= 2e^{2\pi nt}e^{(\pi/2)|\mathrm{Im}(\lambda+1)|}\left\{t^{-\mathrm{Re}(\lambda+1)} - \frac{1}{\mathrm{Re}\,\lambda}\right\}.\end{aligned}$$

Since $n < 0$ and $\operatorname{Re}\lambda > 0$, the latter expression tends to 0 as $t \to +\infty$ and it follows from (9) that

$$\lim_{t \to +\infty} \int_{t-i\infty}^{t+i\infty} z^{-\lambda} e^{2\pi n z}\, dz = 0.$$

This fact together with (8) yields

$$\int_{t-i\infty}^{t+i\infty} z^{-\lambda} e^{2\pi n z}\, dz = 0,$$

for all $t > 0$.

(b) Let $t > 0$ and $\operatorname{Re}\lambda > 1$. Then for arbitrary positive M and N we have

$$\int_{t-iM}^{t+iN} z^{-\lambda}\, dz = (1-\lambda)^{-1}\{(t+iN)^{1-\lambda} - (t-iM)^{1-\lambda}\} \to 0,$$

as $M, N \to +\infty$.

The proof is complete.

Proof of Theorem 4. Write $\tau = \sigma + it, t > 0$, and let $f_t(x) = (x+it)^{-\lambda}$, a continuous function of the real variable x. By Poisson's sum formula (Theorem 7 of Chapter 3),

$$\sum_{m=-\infty}^{\infty} (m+\tau)^{-\lambda} = \sum_{m=-\infty}^{\infty} f_t(\sigma+m) = \sum_{n=-\infty}^{\infty} e^{-2\pi i n\sigma} \int_{-\infty}^{\infty} (x+it)^{-\lambda} e^{2\pi i n x}\, dx,$$

for any σ and t such that the right-hand side converges. The convergence for $t > 0$ and all σ will follow as a consequence of the ensuing calculations.

We have

$$\sum_{m=-\infty}^{\infty} (m+\tau)^{-\lambda} = \sum_{n=-\infty}^{\infty} e^{-2\pi i n\tau} \int_{-\infty}^{\infty} (x+it)^{-\lambda} e^{2\pi i n(x+it)}\, dx.$$

But

$$\int_{-\infty}^{\infty} (x+it)^{-\lambda} e^{2\pi i n(x+it)}\, dx = \int_{-\infty+it}^{\infty+it} u^{-\lambda} e^{2\pi i n u}\, du$$

$$= \frac{-i}{(i)^{\lambda}} \int_{t-i\infty}^{t+i\infty} v^{-\lambda} e^{-2\pi n v}\, dv = 0 \qquad \text{if } n \geqslant 0,$$

by Lemma 6. It follows that

$$\sum_{m=-\infty}^{\infty} (m+\tau)^{-\lambda} = \sum_{n=1}^{\infty} e^{2\pi i n\tau} \frac{(-i)}{(i)^{\lambda}} \int_{t-i\infty}^{t+i\infty} v^{-\lambda} e^{2\pi n v}\, dv.$$

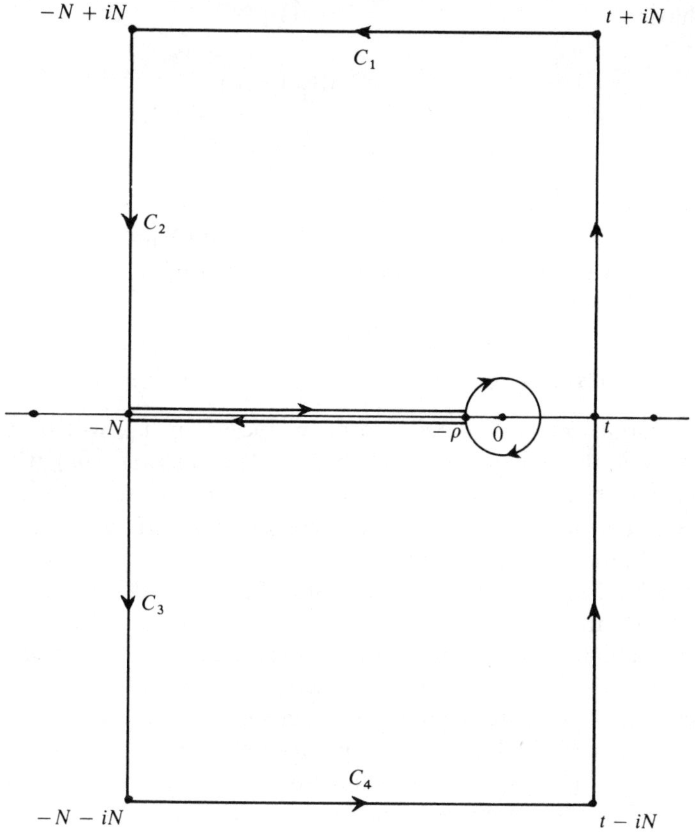

Figure 4. The contour C_N.

We now claim that, for $n > 0$,

$$\int_{t-i\infty}^{t+i\infty} v^{-\lambda} e^{2\pi n v}\, dv = \lim_{N\to+\infty} \int_{D_{N,\rho}} v^{-\lambda} e^{2\pi n v}\, dv, \tag{10}$$

where $t > \rho > 0$ and $D_{N,\rho}$ is the contour described in Lemma 5 (Figure 3). To prove (10) let $N > \rho$ and let C_N be the contour shown in Figure 4. By Cauchy's theorem $\int_{C_N} v^{-\lambda} e^{2\pi n v}\, dv = 0$. On the other hand, simple estimates show that

$$\lim_{N\to+\infty} \int_{C_i} v^{-\lambda} e^{2\pi n v}\, dv = 0, \qquad \text{for } i = 1,2,3,4$$

(see Figure 4). Since the existence of $\int_{t-i\infty}^{t+i\infty} v^{-\lambda} e^{2\pi n v}\, dv$, for any integer n, was proved in Lemma 6(a), (10) follows.

We have

$$\sum_{m=-\infty}^{\infty} (m+\tau)^{-\lambda} = \sum_{n=1}^{\infty} e^{2\pi i n\tau}(i)^{-\lambda}(-i)\int_{t-i\infty}^{t+i\infty} v^{-\lambda}e^{2\pi n v}\,dv$$
$$= \sum_{n=1}^{\infty} e^{2\pi i n\tau}(i)^{-\lambda}\frac{(2\pi)^{\lambda}n^{\lambda-1}}{\Gamma(\lambda)},$$

by (10) and Lemma 5. The convergence of the right-hand side is now clear for $t > 0$. If we replace τ by $\tau/2$ and rearrange slightly we obtain Theorem 4 in the stated form.

3. THE FUNCTION $\psi_s(\tau)$

The function $\psi_s(\tau)$, a modular form on the group Γ_9, will be of great importance in the remainder of the chapter. The function $\rho_s(m)$, which was introduced without motivation in Chapter 5, Section 1, arises rather naturally as the mth coefficient in the Fourier expansion of $\psi_s(\tau)$ at ∞. Before we give the definition of $\psi_s(\tau)$ we need

Lemma 7. If $s \geqslant 5$, then the double sum $\Sigma_c \Sigma'_d |c\tau + d|^{-s/2}$ converges for all $\tau \in \mathscr{H}$. Furthermore, the convergence is uniform on compact subsets of $\mathscr{H}$. The summation here is over all pairs of integers (c,d) except $(c,d) = (0,0)$.

Proof. Let $\tau \in \mathscr{H}$ be fixed and consider the lattice $\{c\tau + d | c$ and d are integers$\}$. Let r be a fixed nonnegative integer and put $\Pi_r = \{\pm r\tau + n;\ n\tau + r | -r \leqslant n \leqslant r\}$. Π_r contains $8r$ points. Let h be the minimum distance of Π_1 to the origin; clearly h is a function of τ. Then rh is the minimum distance of Π_r to the origin, so that, if $c\tau + d \in \Pi_r$, $|c\tau + d| \geqslant rh$. Hence

$$\sum_{c\tau+d\in\Pi_r} |c\tau + d|^{-s/2} \leqslant \frac{8r}{(hr)^{s/2}} = 8h^{-s/2}r^{-s/2+1}.$$

We have

$$\sum_c\sum_d{}' |c\tau + d|^{-s/2} = \sum_{r=1}^{\infty}\sum_{c\tau+d\in\Pi_r} |c\tau + d|^{-s/2}$$

(11)

$$\leqslant 8h^{-s/2}\sum_{r=1}^{\infty} r^{-s/2+1} < \infty,$$

since $s/2 - 1 > 1$.

The estimate in (11) also shows that the convergence is uniform on compact subsets of $\mathscr{H}$.

Definition. Let s be an integer $\geqslant 5$ and put

$$\psi_s(\tau) = \tfrac{1}{2}\sum_c \sum_d{}^* \bar{v}_s(M_{c,d})(c\tau + d)^{-s/2},$$

for $\tau \in \mathscr{H}$. As before, we are summing over all pairs of integers c, d such that $(c,d) = 1$ and c and d of opposite parity. As before, $M_{c,d}$ is any element of Γ_ϑ with lower row c, d. Note that by Lemmas 1 and 7 the function $\psi_s(\tau)$ is well defined.

The next result is the first step in proving that $\psi_s(\tau)$ is an entire modular form on the group Γ_ϑ, with multiplier system v_s.

Theorem 8. The function $\psi_s(\tau)$ is regular in $\mathscr{H}$ and there satisfies the functional equations

$$\psi_s(M\tau) = v_s(M)(\gamma\tau + \delta)^{s/2}\psi_s(\tau),$$

for all $M = \begin{pmatrix} \alpha & \beta \\ \gamma & \delta \end{pmatrix} \in \Gamma_\vartheta$.

Proof. That $\psi_s(\tau)$ is regular in $\mathscr{H}$ is clear since by Lemma 7 the series defining $\psi_s(\tau)$ converges absolutely uniformly on all compact subsets of $\mathscr{H}$.

By definition of $\psi_s(\tau)$ we have

$$\psi_s(M\tau) = \tfrac{1}{2}\sum_c \sum_d{}^* \bar{v}_s(M_{c,d})(cM\tau + d)^{-s/2}.$$

By the consistency condition (1) we have

$$v_s(M_{c,d})(cM\tau + d)^{s/2} = \frac{v_s(M_{c,d}M)(c'\tau + d')^{s/2}}{v_s(M)(\gamma\tau + \delta)^{s/2}},$$

where

$$M_{c,d}M = \begin{pmatrix} * & * \\ c' & d' \end{pmatrix} = \begin{pmatrix} * & * \\ \alpha c + \gamma d & \beta c + \delta d \end{pmatrix},$$

so that

$$\begin{aligned} \psi_s(M\tau) &= \tfrac{1}{2}v_s(M)(\gamma\tau + \delta)^{s/2}\sum_c \sum_d{}^* \bar{v}_s(M_{c,d}M)(c'\tau + d')^{-s/2} \\ &= \tfrac{1}{2}v_s(M)(\gamma\tau + \delta)^{s/2}\sum_c \sum_d{}^* \bar{v}_s(M_{c',d'})(c'\tau + d')^{-s/2}. \end{aligned}$$

Now it is a simple matter to verify that as c, d runs through all pairs of relatively prime integers of opposite parity so does $c' = \alpha c + \gamma d, d' = \beta c + \delta d$. Thus by absolute convergence of the series defining $\psi_s(\tau)$ it follows that

$$\tfrac{1}{2}\sum_c \sum_d{}^* \bar{v}_s(M_{c',d'})(c'\tau + d')^{-s/2} = \psi_s(\tau).$$

and we conclude that

$$\psi_s(M\tau) = v_s(M)(\gamma\tau + \delta)^{s/2}\psi_s(\tau).$$

The proof is complete.

By Example 1 of Chapter 2, Section 1, there is a S.F.R. of Γ_ϑ in which the parabolic points are ∞ and -1. With an eye toward proving that $\psi_s(\tau)$ is an entire modular form on Γ_ϑ, we therefore examine its expansions at ∞ and -1. We have

Theorem 9. The Fourier expansions of $\psi_s(\tau)$ at ∞ is

$$\psi_s(\tau) = 1 + \sum_{m=1}^{\infty} \rho_s(m)e^{\pi i m\tau},$$

where $\rho_s(m)$ is defined by (5).

Proof. By (3) and the fact that $v_s(M_{0,1}) = v_s(S^2) = 1$, we have

$$\psi_s(\tau) = 1 + \sum_{c>0} \sum_{d}{}^{*} \bar{v}_s(M_{c,d})(c\tau + d)^{-s/2}.$$

We shall rewrite the inner sum on d. We can write d *uniquely* in the form $d = h + 2mc$, where m is an integer and $0 \leqslant h < 2c$. Since d is relatively prime to c and of parity opposite to that of c, the same is true of h. Thus we can write

$$M_{c,h} = M_{c,d}(S^2)^{-m},$$

since $h = d - 2mc$. Of course, the upper rows of $M_{c,h}$ and $M_{c,d}$ have to be chosen appropriately. By (2) we have

$v_s(M_{c,h}) = v_s(M_{c,d})$, so that

$$\psi_s(\tau) = 1 + \sum_{c>0} \sum_{0\leqslant h<2c}{}^{*} \bar{v}_s(M_{c,h}) \sum_{m=-\infty}^{\infty} (c\tau + h + 2mc)^{-s/2}.$$

Note that we have rearranged the double sum. This is justified by the absolute convergence. By Theorem 4, with τ replaced by $\tau + h/c$, we have

$$\sum_{m=-\infty}^{\infty} (2m + \tau + h/c)^{-s/2} = \frac{e^{-\pi i s/4}\pi^{s/2}}{\Gamma(s/2)} \sum_{n=1}^{\infty} n^{s/2-1}e^{\pi i n(\tau + h/c)}.$$

Thus

$$(12) \quad \begin{aligned} \psi_s(\tau) &= 1 + \sum_{c>0} c^{-s/2} \sum_{0\leqslant h<2c}{}^{*} \bar{v}_s(M_{c,h}) \sum_{m=-\infty}^{\infty} (2m + \tau + h/c)^{-s/2} \\ &= 1 + \frac{e^{-\pi i s/4}\pi^{s/2}}{\Gamma(s/2)} \sum_{c>0} c^{-s/2} \sum_{0\leqslant h<2c}{}^{*} \bar{v}_s(M_{c,h}) \sum_{n=1}^{\infty} n^{s/2-1}e^{\pi i n(\tau + h/c)}. \end{aligned}$$

Since $s \geqslant 5$ the double sum on the right-hand side of (12) is absolutely convergent. Thus we may rearrange it to complete the proof of the theorem.

4. THE EXPANSION OF $\psi_s(\tau)$ AT -1

We need a preliminary lemma.

Lemma 10. Suppose c and d are integers such that $c > 0$, d is odd, and $(c,d) = 1$. Then the following hold.

(a) If c is even, then

$$\left(\frac{c}{|c+d|}\right) = \left(\frac{c}{|d|}\right)(-1)^{(d+1)c/4}.$$

(b)

$$\bar{\upsilon}_\vartheta(M_{c,d+c}) = \left(\frac{c}{|d|}\right)e^{(\pi i/4)d(c-1)}e^{\pi i/4}.$$

(c)

$$\frac{\left(\frac{-c}{|d|}\right)e^{-(\pi i d/4)(-c-1)}}{(-d\tau+c)^{1/2}} = -\frac{\left(\frac{c}{|d|}\right)e^{(\pi i d/4)(c-1)}}{(d\tau-c)^{1/2}}.$$

Proof. Throughout the proof of Lemma 10 we make extensive use, without comment, of Lemma 1 of Chapter 4.

(a) Put $c = 2^q c_1$, where $q \geqslant 1$ and c_1 is odd. Then

$$\left(\frac{c}{|d|}\right) = \left(\frac{2}{|d|}\right)^q\left(\frac{c_1}{|d|}\right) = (-1)^{\frac{d^2-1}{8}q}\left(\frac{d}{c_1}\right)(-1)^{\frac{d-1}{2}\frac{c_1-1}{2}}.$$

If $q = 1$, then

$$\left(\frac{c}{|c+d|}\right) = \left(\frac{2}{|c+d|}\right)\left(\frac{c_1}{|c+d|}\right) = (-1)^{\frac{(c+d)^2-1}{8}}\left(\frac{c+d}{c_1}\right)(-1)^{\frac{c+d-1}{2}\frac{c_1-1}{2}}$$

$$= (-1)^{\frac{d^2-1}{8}}\left(\frac{d}{c_1}\right)(-1)^{\frac{d-1}{2}\frac{c_1-1}{2}}(-1)^{\frac{c(2d+c)}{8}}(-1)^{\frac{c}{2}\frac{c_1-1}{2}}$$

$$= \left(\frac{c}{|d|}\right)(-1)^{\frac{c_1(d+c_1)}{2}}(-1)^{\frac{c_1(c_1-1)}{2}}$$

$$= \left(\frac{c}{|d|}\right)(-1)^{\frac{c_1}{2}(d+2c_1-1)} = \left(\frac{c}{|d|}\right)(-1)^{\frac{c_1 d}{2}+c_1^2-\frac{c_1}{2}}$$

$$= \left(\frac{c}{|d|}\right)(-1)^{\frac{c_1}{2}(d-1)+c_1} = \left(\frac{c}{|d|}\right)(-1)^{\frac{c_1}{2}(d+1)} = \left(\frac{c}{|d|}\right)(-1)^{\frac{(d+1)c}{4}},$$

the stated result.

If $q > 1$, then $4|c$ and the statement becomes $\left(\dfrac{c}{|c+d|}\right) = \left(\dfrac{c}{|d|}\right)$. As before we get

$$\left(\frac{c}{|c+d|}\right) = \left(\frac{c}{|d|}\right)(-1)^{\frac{qc(2d+c)}{8}}(-1)^{\frac{(c/2)(c_1-1)}{2}} = \left(\frac{c}{|d|}\right)(-1)^{\frac{qcd}{4}}.$$

Now if $q \geqslant 2$, then $8|qcd$, so that $(-1)^{qcd/4} = 1$, and we obtain the desired result.

(b) Since $(c,d) = 1$ and d is odd, it follows that $(c, d + c) = 1$ and $c, d + c$ are of opposite parity. Thus there exists $M_{c,d+c} \in \Gamma_\vartheta$, an element of Γ_ϑ with lower row $c, d + c$ [see Lemma 1(a)]. By Theorem 3 of Chapter 4 we have, since $c > 0$,

$$\bar{v}_\vartheta(M_{c,d+c}) = \begin{cases} \left(\dfrac{d+c}{c}\right)e^{\pi ic/4} & \text{if } c \text{ is odd,} \\[2ex] \left(\dfrac{c}{|d+c|}\right)e^{-\pi i(c+d-1)/4} & \text{if } c \text{ is even.} \end{cases}$$

If c is odd we have

$$\bar{v}_\vartheta(M_{c,d+c}) = \left(\frac{d}{c}\right)e^{\pi ic/4} = \left(\frac{c}{|d|}\right)(-1)^{\frac{c-1}{2}\frac{d-1}{2}}e^{\pi ic/4}$$

$$= \left(\frac{c}{|d|}\right)e^{(\pi i/4)\{(c-1)(d-1)+c\}} = \left(\frac{c}{|d|}\right)e^{(\pi i/4)d(c-1)}e^{\pi i/4}.$$

If c is even,

$$\bar{v}_\vartheta(M_{c,d+c}) = \left(\frac{c}{|d|}\right)(-1)^{\frac{d+1}{4}}e^{-\pi i(c+d-1)/4} = \left(\frac{c}{|d|}\right)e^{(\pi i/4)d(c-1)}e^{\pi i/4},$$

where we have used part (a).

(c) Suppose $d > 0$. Then by our branch convention $(-d\tau + c)^{1/2} = e^{-\pi i/2}(d\tau - c)^{1/2}$. Hence

$$\frac{\left(\dfrac{-c}{|d|}\right)e^{(-\pi id/4)(-c-1)}}{(-d\tau+c)^{1/2}} = \frac{(-1)^{\frac{d-1}{2}}\left(\dfrac{c}{|d|}\right)e^{(\pi id/4)(c+1)}}{e^{-\pi i/2}(d\tau-c)^{1/2}}$$

$$= \frac{\left(\dfrac{c}{|d|}\right)e^{(\pi id/4)(c-1)}e^{\pi id}}{(d\tau-c)^{1/2}} = -\frac{\left(\dfrac{c}{|d|}\right)e^{(\pi id/4)(c-1)}}{(d\tau-c)^{1/2}}.$$

If $d < 0$, the branch convention gives $(-d\tau + c)^{1/2} = e^{\pi i/2}(d\tau - c)^{1/2}$. Thus

$$\frac{\left(\frac{-c}{|d|}\right)e^{(-\pi id/4)(-c-1)}}{(-d\tau + c)^{1/2}} = \frac{(-1)^{\frac{-d-1}{2}}\left(\frac{c}{|d|}\right)e^{(\pi id/4)(c+1)}}{e^{\pi i/2}(d\tau - c)^{1/2}}$$

$$= \frac{\left(\frac{c}{|d|}\right)e^{\pi id/2}e^{\pi idc/4}e^{\pi id/4}}{(d\tau - c)^{1/2}} = \frac{\left(\frac{c}{|d|}\right)e^{(\pi id/4)(c-1)}e^{\pi id}}{(d\tau - c)^{1/2}}$$

$$= -\frac{\left(\frac{c}{|d|}\right)e^{(\pi id/4)(c-1)}}{(d\tau - c)^{1/2}},$$

and the proof of Lemma 10 is complete.

We now turn to the expansion of $\psi_s(\tau)$ at the parabolic point -1. Recall that in the proof of Theorem 9 we had

$$\psi_s(\tau) = 1 + \sum_{c>0} \sum_{d}{}^{*} \bar{v}_s(M_{c,d})(c\tau + d)^{-s/2},$$

for $\tau \in \mathscr{H}$. From this it follows that

$$\psi_s(-1 - 1/\tau) = 1 + \sum_{c>0} \sum_{d}{}^{*} \bar{v}_s(M_{c,d})\left(\frac{-c}{\tau} + d - c\right)^{-s/2}.$$

By our branch convention

$$\tau^{1/2}\left(\frac{-c}{\tau} + d - c\right)^{1/2} = (-1)^{\frac{\operatorname{sign}(d-c)-1}{2}}(d\tau - c\tau - c)^{1/2},$$

so that,

$$\psi_s(-1 - 1/\tau) = 1 + \tau^{s/2} \sum_{c>0} \sum_{d}{}^{*} \frac{(-1)^{\frac{s}{2}\{\operatorname{sign}(d-c)-1\}}\bar{v}_s(M_{c,d})}{(d\tau - c\tau - c)^{s/2}}$$

$$= 1 + \tau^{s/2} \sum_{c>0} \sum_{\substack{d' \\ (c,d')=1 \\ d' \text{ odd}}} \frac{(-1)^{\frac{\operatorname{sign} d'-1}{2}s}\bar{v}_s(M_{c,d'+c})}{(d'\tau - c)^{s/2}}$$

$$= 1 + e^{\pi is/4}\tau^{s/2} \sum_{c>0} \sum_{\substack{d \\ (c,d)=1 \\ d \text{ odd}}} \frac{\left\{(-1)^{\frac{\operatorname{sign} d-1}{2}}\left(\frac{c}{|d|}\right)e^{(\pi id/4)(c-1)}\right\}^s}{(d\tau - c)^{s/2}}.$$

Here we have replaced $d - c$ by d', used the absolute convergence of the double sum, and applied Lemma 10(b).

We now consider those pairs c, d in the summation with $d < 0$. We replace these by $-c, -d$; this gives a new pair c, d with $c < 0$ and $d > 0$. Using Lemma 10(c) we obtain

$$\psi_s(-1 - 1/\tau) = 1 + e^{\pi i s/4}\tau^{s/2} \sum_{\substack{c \neq 0 \\ (c,d)=1 \\ d \text{ odd}}} \sum_{d > 0} \frac{\left\{\left(\frac{c}{|d|}\right) e^{(\pi i d/4)(c-1)}\right\}^s}{(d\tau - c)^{s/2}}$$

$$= e^{\pi i s/4}\tau^{s/2} \sum_{\substack{d > 0 \\ d \text{ odd}}} \sum_{\substack{c=-\infty \\ (c,d)=1}}^{\infty} \frac{\left\{\left(\frac{c}{d}\right) e^{(\pi i d/4)(c-1)}\right\}^s}{(d\tau - c)^{s/2}},$$

where we have absorbed 1 into the sum as the term corresponding to the pair $c = 0$, $d = 1$, and used the absolute convergence of the double sum to interchange the summations on c and d. Next we make the substitution $c = l + 8dm$, subject to the conditions $(l,d) = 1$ and $0 \leqslant l < 8d$. A simple consideration shows that as l takes on all integer values satisfying these two conditions and m runs through all the integers, $l + 8dm$ takes on exactly once each integer value c such that $(c,d) = 1$. Thus, applying absolute convergence once more, we obtain

$$\psi_s(-1-1/\tau) = e^{\pi i s/4}\tau^{s/2} \sum_{\substack{d>0 \\ d \text{ odd}}} \sum_{\substack{0 \leqslant l < 8d \\ (l,d)=1}} \sum_{m=-\infty}^{\infty} \frac{\left(\frac{l+8dm}{d}\right)^s e^{(\pi i d/4)(l+8dm-1)s}}{(d\tau - l - 8dm)^{s/2}}$$

$$= 4^{-s/2} e^{\pi i s/4}\tau^{s/2} \sum_{\substack{d>0 \\ d \text{ odd}}} e^{-\pi i d s/4} d^{-s/2} \sum_{\substack{0 \leqslant l < 8d \\ (l,d)=1}} \left(\frac{l}{d}\right)^s e^{\pi i d l s/4}$$
$$\times \sum_{m=-\infty}^{\infty} \left(-2m + \frac{\tau}{4} - \frac{l}{4d}\right)^{-s/2}.$$

An application of the Lipschitz summation formula (Theorem 4) to the inner sum on m yields

$$\psi_s(-1 - 1/\tau) = \frac{\pi^{s/2}\tau^{s/2}}{4^{s/2}\Gamma(s/2)} \sum_{\substack{d>0 \\ d \text{ odd}}} e^{-\pi i d s/4} d^{-s/2} \sum_{\substack{0 \leqslant l < 8d \\ (l,d)=1}} \left(\frac{l}{d}\right)^s e^{\pi i d l s/4}$$

(13)

$$\times \sum_{n=1}^{\infty} n^{s/2-1} \exp\left[\pi i n\left(\frac{\tau}{4} - \frac{l}{4d}\right)\right].$$

The multiple sum on the right side of (13) is absolutely convergent since $\tau \in \mathscr{H}$ and $s \geqslant 5$. Thus the right-hand side of (13) can be rearranged to give, finally,

Theorem 11. For $\tau \in \mathscr{H}$, we have

$$\psi_s(-1 - 1/\tau) = \tau^{s/2} \sum_{n=1}^{\infty} \varphi_s(n) e^{2\pi i n\tau/8},$$

where

$$\varphi_s(n) = \frac{(2\pi)^{s/2}}{\Gamma(s/2)} (n/8)^{s/2-1} \sum_{\substack{d>0 \\ d \text{ odd}}} D_d(n),$$

with

$$D_d(n) = \tfrac{1}{8} d^{-s/2} e^{-\pi i d s/4} \sum_{\substack{0 \leqslant l < 8d \\ (l,d)=1}} \left(\frac{l}{d}\right)^s e^{(2\pi i l/8d)(sd^2 - n)}.$$

Remark. In Theorem 11 we replace τ by $-1/(\tau + 1)$ and use the fact that $(-\tau - 1)^{s/2} = (-1)^{s/2}(\tau + 1)^{s/2}$. This gives

$$(14) \qquad \psi_s(\tau) = (\tau + 1)^{-s/2} \sum_{n=1}^{\infty} (-1)^{-s/2} \varphi_s(n) \exp\left\{\frac{2\pi i n}{8}[-1/(\tau + 1)]\right\}.$$

Equation (14) has almost the appearance of the Fourier expansion, at the parabolic cusp -1, of a modular form of degree $-s/2$ on the group Γ_ϑ. With a little more effort we can rearrange (14) into precisely such an expansion. For this purpose we shall apply Theorem 4 of Chapter 2, which, by virtue of Theorem 8, implies that $\psi_s(\tau)$ has an expansion at -1 of the form

$$(15) \qquad \psi_s(\tau) = (\tau + 1)^{-s/2} \sum_{n=-\infty}^{\infty} a_s(n) \exp\left\{\frac{2\pi i(n + \kappa)}{\lambda}\left(-\frac{1}{\tau + 1}\right)\right\},$$

valid in all of $\mathscr{H}$. Here λ is the width of the cusp -1, that is, the minimal positive integer such that

$$\begin{pmatrix} 1 & 1 \\ -1 & 0 \end{pmatrix}\begin{pmatrix} 1 & \lambda \\ 0 & 1 \end{pmatrix}\begin{pmatrix} 0 & -1 \\ 1 & 1 \end{pmatrix} = \begin{pmatrix} 1 + \lambda & \lambda \\ -\lambda & 1 - \lambda \end{pmatrix} \in \Gamma_\vartheta,$$

and κ is defined by

$$v_s\left(\begin{pmatrix} 1 + \lambda & \lambda \\ -\lambda & 1 - \lambda \end{pmatrix}\right) = e^{2\pi i \kappa}, \qquad 0 \leqslant \kappa < 1.$$

It is clear from the characterization of Γ_ϑ given by Corollary 4 of Chapter 1 that $\lambda = 1$, so that κ is defined by

$$v_s\left(\begin{pmatrix} 2 & 1 \\ -1 & 0 \end{pmatrix}\right) = e^{2\pi i\kappa}, \qquad 0 \leqslant \kappa < 1.$$

By Theorem 3 of Chapter 4, we have

$$v_\vartheta\left(\begin{pmatrix} 2 & 1 \\ -1 & 0 \end{pmatrix}\right) = e^{2\pi i/8},$$

so that $\kappa = s/8 - [s/8] = \{s/8\}$, the fractional part of $s/8$. (The values of λ and κ could also have been read off from Theorem 13 of Chapter 3.) Thus (15) assumes the form

$$(16) \quad \psi_s(\tau) = (\tau + 1)^{-s/2} \sum_{n=-\infty}^{\infty} a_s(n) \exp\left[2\pi i\left(n + \left\{\frac{s}{8}\right\}\right)\left(-\frac{1}{\tau + 1}\right)\right].$$

Comparing (14) with (16) and using the uniqueness of the Laurent expansion in a given annulus, we conclude that $a_s(n) = 0$ for $n + \{s/8\} \leqslant 0$ and $\varphi_s(n) = 0$ unless $n \equiv \mu \pmod 8$, where $\{s/8\} = \mu/8$, $0 \leqslant \mu < 8$. Since $\mu \equiv s \pmod 8$ we can state

Corollary 12. With $\varphi_s(n)$ as defined in Theorem 11, we have $\varphi_s(n) = 0$ unless $n \equiv s \pmod 8$. It follows that $\psi_s(\tau)$ has the expansion at -1,

$$(17) \quad \psi_s(\tau) = (\tau + 1)^{-s/2} \sum_{\substack{n=1 \\ n \equiv s \pmod 8}}^{\infty} (-1)^{-s/2}\varphi_s(n) \exp\left\{\frac{2\pi i n}{8}[-1/(\tau + 1)]\right\},$$

valid in all of $\mathscr{H}$. The expansion can also be written in the form

$$(18) \quad \psi_s(\tau) = (\tau + 1)^{-s/2} \sum_{n + \{s/8\} > 0} a_s(n) \exp\left[2\pi i\left(n + \left\{\frac{s}{8}\right\}\right)\left(\frac{-1}{\tau + 1}\right)\right].$$

From Theorems 8 and 9 and Corollary 12 we immediately derive

Corollary 13. $\psi_s(\tau)$ is an entire modular form of degree $-s/2$ with multiplier system v_s, on the group Γ_ϑ.

5. PROOFS OF THEOREMS 2 AND 3

We are now in a position to prove the main results of this chapter.

Proof of Theorem 2. By the definition of $r_s(m)$ given in Chapter 3, Section 4, we have

$$(19) \qquad \vartheta^s(\tau) = 1 + \sum_{m=1}^{\infty} r_s(m) e^{\pi i m\tau},$$

valid in all of $\mathscr{H}$. With $s \in Z$ we also observe from Theorem 13 of Chapter 3 that $\vartheta^s(\tau)$ is an entire modular form of degree $-s/2$ with multiplier system $v_\vartheta^s = v_s$ on Γ_ϑ. Equation (19) is of course the expansion of $\vartheta^s(\tau)$ at ∞. Theorem 13 of Chapter 3 also gives the expansion of $\vartheta(\tau)$ at the cusp -1. From this expansion we obtain the following expansion for $\vartheta^s(\tau)$ at -1:

$$\vartheta^s(\tau) = (\tau + 1)^{-s/2} \sum_{m \geqslant 0} b_m e^{2\pi i(m+s/8)[-1/(\tau+1)]}, \qquad b_0 \neq 0. \tag{20}$$

From Theorem 9 and equations (18), (19), and (20) we conclude that for $s \geqslant 5$, $\vartheta^s(\tau) - \psi_s(\tau)$ is a *cusp form* of degree $-s/2$ with multiplier system v_s on Γ_ϑ. Since the mth coefficient of $\vartheta^s(\tau) - \psi_s(\tau)$ in the expansion at ∞ is $r_s(m) - \rho_s(m)$, it follows from Theorem 10 of Chapter 2 that

$$r_s(m) = \rho_s(m) + O(m^{s/4}) \qquad \text{as } m \to +\infty.$$

The proof is complete.

Proof of Theorem 3. Instead of considering $\vartheta^s(\tau) - \psi_s(\tau)$ as we did in the proof of Theorem 2, we here form the quotient $\Phi_s(\tau) = \psi_s(\tau)/\vartheta^s(\tau)$ for $s \geqslant 5$. From the known transformation properties and expansions of $\psi_s(\tau)$ and $\vartheta^s(\tau)$ we conclude that $\Phi_s(\tau)$ is a modular *function* on Γ_ϑ (i.e., a modular form of degree $r = 0$ and multiplier system $v \equiv 1$) with the expansions

$$\text{(21)} \quad \begin{cases} \Phi_s(\tau) = 1 + \displaystyle\sum_{m=1}^{\infty} c_m e^{\pi i m \tau} & \text{at } \infty, \\ \displaystyle\sum_{m+\{s/8\}>0} d_m \exp\left[2\pi i\left(m + \left\{\frac{s}{8}\right\} - \frac{s}{8}\right)\left(\frac{-1}{\tau+1}\right)\right] & \text{at } -1. \end{cases}$$

We suppose now that $5 \leqslant s \leqslant 8$. For $5 \leqslant s \leqslant 7$, we have $\{s/8\} = s/8$, so that the first term in the expansion at -1 is d_0. If $s = 8$, then $\{s/8\} = 0$, but the summation condition on m is "$m > 0$." Hence the first term in the expansion at -1 is d_1 in this case. In either case the expansion has no negative powers of $\exp\left\{2\pi i\left(\dfrac{-1}{\tau+1}\right)\right\}$.

On the other hand, since $\vartheta(\tau) \neq 0$ in $\mathscr{H}$ it follows that $\Phi_s(\tau)$ is regular in $\mathscr{H}$. Theorem 7 of Chapter 2 thus implies that $\Phi_s(\tau)$ is a constant. Since $\Phi_s(\tau) \to 1$ as $\tau \to i\infty$, it follows that $\Phi_s(\tau) \equiv 1$ for $5 \leqslant s \leqslant 8$. Thus for $5 \leqslant s \leqslant 8$ we have $\vartheta^s(\tau) = \psi_s(\tau)$, from which it follows that $r_s(m) = \rho_s(m)$ for $m \geqslant 1$. Note that the proof makes essential use of the fact that $s \leqslant 8$. The theorem is, in fact, false for $s \geqslant 9$, as we shall prove in Section 6.

6. RELATED RESULTS

In Theorem 2 we have shown that $r_s(m) = \rho_s(m) + O(m^{s/4})$, as $m \to +\infty$. In order to show that this asymptotic estimate is significant we must show that $\rho_s(m)$ is of greater order of growth than the error term. This is part of the content of our next theorem. Note that by Theorem 3 the error term is identically zero for $5 \leqslant s \leqslant 8$. Hence it suffices to prove the order result for $s \geqslant 9$.

Theorem 14. For $s \geqslant 5$, $\rho_s(n)$ is real, for all n. Furthermore, for $s \geqslant 6$ there exist positive constants K_1, K_2, independent of n such that

$$K_1 n^{s/2-1} < \rho_s(n) < K_2 n^{s/2-1}, \qquad \text{for } n = 1, 2, \ldots.$$

Since for $s \geqslant 5$ we have $s/4 < s/2 - 1$, $\rho_s(n)$ is of greater order of growth than the error term $O(n^{s/4})$.

In the proof of Theorem 14 we need

Lemma 15. If c and h are positive integers with c even, h odd, and $(c,h) = 1$, then $\left(\dfrac{c}{2c-h}\right) = \varepsilon\left(\dfrac{c}{h}\right)$, where $\varepsilon = 1$ if $c \equiv 0 \pmod 4$, and $\varepsilon = -1$ if $c \equiv 2 \pmod 4$.

Proof. Write $c = 2^\alpha c_1$, with $\alpha \geqslant 1$ and c_1 odd. Then using Lemma 1 of Chapter 4 we have

$$\begin{aligned}
\left(\frac{c}{2c-h}\right) = \left(\frac{c}{2c-h}\right)^{\alpha}\left(\frac{c_1}{2c-h}\right) &= (-1)^{\frac{(2c-h)^2-1}{8}\alpha}\left(\frac{2c-h}{c_1}\right)(-1)^{\frac{c_1-1}{2}\frac{2c-h-1}{2}} \\
&= (-1)^{\frac{h^2-1}{8}\alpha}\left(\frac{-h}{c_1}\right)(-1)^{\frac{c_1-1}{2}\frac{2c-h-1}{2}}(-1)^{\frac{-4ch\alpha}{8}} \\
&= \left(\frac{2}{h}\right)^{\alpha}\left(\frac{h}{c_1}\right)(-1)^{\frac{c_1-1}{2}\frac{1-h}{2}}\,\varepsilon \\
&= \left(\frac{2}{h}\right)^{\alpha}\left(\frac{c_1}{h}\right)(-1)^{\frac{c_1-1}{2}\frac{h-1}{2}}(-1)^{\frac{c_1-1}{2}\frac{1-h}{2}}\,\varepsilon \\
&= \varepsilon\left(\frac{c}{h}\right).
\end{aligned}$$

We now return to the

Proof of Theorem 14. We shall show first that $\rho_s(n)$ is real for all n. Recall that

$$\rho_s(n) = \frac{e^{-\pi i s/4}\pi^{s/2}}{\Gamma(s/2)} n^{s/2-1} \sum_{c=1}^{\infty} B_c(n),$$

where

$$B_c(n) = c^{-s/2} \sum_{0 \leqslant h < 2c}^{*} \bar{v}_s(M_{c,h}) e^{\pi i n h/c}.$$

We now utilize Theorem 3 of Chapter 4. If c is odd, then since $c > 0$ and $h \geqslant 0$,

$$\bar{v}_s(M_{c,h}) = \left(\frac{h}{c}\right)^s e^{\pi isc/4}.$$

Hence for odd c, we have

$$\overline{e^{-\pi is/4}B_c(n)} = e^{\pi is/4}c^{-s/2} \sum_{0 \leqslant h < 2c}^{*} \left(\frac{h}{c}\right)^s e^{-\pi isc/4}e^{-\pi inh/c}.$$

Replacing h by $2c - h$, we obtain

$$c^{-s/2}e^{(\pi is/4)(1-c)}(-1)^{\frac{c-1}{2}s} \sum_{0 \leqslant h < 2c}^{*} \left(\frac{h}{c}\right)^s e^{\pi inh/c}$$

$$= c^{-s/2}e^{(\pi is/4)(c-1)} \sum_{0 \leqslant h < 2c}^{*} \left(\frac{h}{c}\right)^s e^{\pi inh/c}$$

$$= c^{-s/2}e^{-\pi is/4} \sum_{0 \leqslant h < 2c}^{*} \left(\frac{h}{c}\right)^s e^{\pi ics/4}e^{\pi inh/c} = e^{-\pi is/4}B_c(n).$$

If c is even we have, again by Theorem 3 of Chapter 4,

$$\bar{v}_s(M_{c,h}) = \left(\frac{c}{h}\right)^s e^{-\pi is(h-1)/4}.$$

Thus for even c,

$$\overline{e^{-\pi is/4}B_c(n)} = e^{\pi is/4}c^{-s/2} \sum_{0 \leqslant h < 2c}^{*} \left(\frac{c}{h}\right)^s e^{\pi i(h-1)s/4}e^{-\pi inh/c}$$

$$= c^{-s/2} \sum_{0 \leqslant h < 2c}^{*} \left(\frac{c}{h}\right)^s e^{\pi ihs/4}e^{-\pi inh/c}.$$

If we again replace h by $2c - h$, this becomes

$$c^{-s/2} \sum_{0 \leqslant h < 2c}^{*} \left(\frac{c}{2c-h}\right)^s e^{\pi i(2c-h)s/4}e^{-\pi in(2c-h)/c},$$

which, by Lemma 15, is in turn equal to

$$c^{-s/2}\varepsilon^s \sum_{0 \leqslant h < 2c}^{*} \left(\frac{c}{h}\right)^s e^{-\pi ihs/4}e^{\pi inh/c}(-1)^{\frac{c}{2}s}$$

$$= c^{-s/2}\varepsilon^{2s} \sum_{0 \leqslant h < 2c}^{*} \left(\frac{c}{h}\right)^s e^{-\pi ihs/4}e^{\pi inh/c} = e^{-\pi is/4}B_c(n).$$

This shows that $e^{-\pi is/4}B_c(n)$ is real for all c and all n. Thus $\rho_s(n)$ is real for all n.

We next turn to the order estimate of $\rho_s(n)$. If $s \geqslant 5$, then from the definition and the fact that $\rho_s(n)$ is real we have

$$\rho_s(n) \leqslant \frac{\pi^{s/2}}{\Gamma(s/2)} n^{s/2-1} \sum_{c=1}^{\infty} c^{-s/2}(2c)$$
$$< K n^{s/2-1} \zeta(s/2 - 1) = K_2 n^{s/2-1},$$

where $K_2 > 0$ is a constant independent of n. Here $\zeta(z) = \sum_{n=1}^{\infty} 1/n^z$, $\mathrm{Re}\, z > 1$, is the well-known Riemann ζ-function.

On the other hand,

$$\rho_s(n) \geqslant \frac{e^{-\pi i s/4}\pi^{s/2}}{\Gamma(s/2)} n^{s/2-1} B_1(n) - \frac{\pi^{s/2}}{\Gamma(s/2)} n^{s/2-1} \sum_{c=2}^{\infty} c^{-s/2+1}.$$

Since $B_1(n) = e^{\pi i s/4}$, this becomes

$$\rho_s(n) \geqslant \frac{\pi^{s/2}}{\Gamma(s/2)} n^{s/2-1} - \frac{\pi^{s/2}}{\Gamma(s/2)} n^{s/2-1}\{\zeta(s/2 - 1) - 1\}$$
$$= \frac{\pi^{s/2}}{\Gamma(s/2)} n^{s/2-1}\{2 - \zeta(s/2 - 1)\}.$$

But it is obvious from the definition of $\zeta(z)$ that it is a monotone decreasing function of z for real $z > 1$. Hence for $s \geqslant 6$, $\zeta(s/2 - 1) \leqslant \zeta(2) = \pi^2/6 < \frac{11}{6}$. Hence for $s \geqslant 6$,

$$\rho_s(n) > \frac{\pi^{s/2}}{\Gamma(s/2)} \frac{1}{6} n^{s/2-1} = K_1 n^{s/2-1}.$$

This completes the proof.

In Theorem 3 we proved that $\vartheta^s(\tau) = \psi_s(\tau)$ if $5 \leqslant s \leqslant 8$. We have also stated without proof that $\vartheta^s(\tau) \neq \psi_s(\tau)$ if $s \geqslant 9$. We conclude this chapter with a proof of this latter fact. We begin with

Lemma 16. Let d be an odd positive integer. Then with $D_d(n)$ defined as in Theorem 11 and $B_d(n)$ defined by (4) we have

$$D_d(n) = \begin{cases} e^{-\pi i s/2} B_d(n) & \text{if } n \equiv s \pmod 8, \\ 0 & \text{if } n \not\equiv s \pmod 8. \end{cases}$$

Remark. Lemma 16 implies that $\varphi_s(n) = 0$ if $n \not\equiv s \pmod 8$, a result already stated in Corollary 12. The proof of Corollary 12 depended upon an examination of the Fourier expansion of $\psi_s(\tau)$ at the cusp -1, whereas the proof of Lemma 16 is purely number-theoretic. It is also clear that Lemma 16 is a stronger result than is Corollary 12.

Proof. With d odd and n a positive integer we had defined $D_d(n)$ by

$$D_d(n) = \tfrac{1}{8}d^{-s/2}e^{-\pi ids/4} \sum_{\substack{0 \leqslant l < 8d \\ (l,d)=1}} \left(\frac{l}{d}\right)^s e^{2\pi il(sd^2-n)/8d}. \tag{22}$$

Since d is odd, we have $(d,8) = 1$, so that every integer l can be written in the form $l = gd + 8j$, with g and j integers. Furthermore, $(gd + 8j, d) = 1$ if and only if $(j,d) = 1$. Suppose we let g range through the integers between 0 and 7 inclusive and j through a *reduced* residue system modulo d. In this way we get $8\varphi(d)$ integers $l = gd + 8j$ such that $(l,d) = 1$. In the range $0 \leqslant l < 8d$ there are *precisely* $8\varphi(d)$ integers l such that $(l,d) = 1$. But it is a simple matter to check that the numbers $gd + 8j$, with $0 \leqslant g \leqslant 7$ and $0 \leqslant j \leqslant d - 1$, $(j,d) = 1$, are all distinct modulo $8d$. Since the terms in the summation occurring in (22) depend only upon the residue class of l modulo $8d$, it follows that

$$D_d(n) = \tfrac{1}{8}d^{-s/2}e^{-\pi ids/4} \sum_{\substack{0 \leqslant j < d \\ (j,d)=1}} \sum_{0 \leqslant g < 8} \left(\frac{gd + 8j}{d}\right)^s e^{2\pi i(gd+8j)(sd^2-n)/8d}$$

$$= \tfrac{1}{8}d^{-s/2}e^{-\pi ids/4} \sum_{\substack{0 \leqslant j < d \\ (j,d)=1}} \left(\frac{8j}{d}\right)^s e^{2\pi ij(sd^2-n)/d} \sum_{0 \leqslant g < 8} e^{2\pi ig(sd^2-n)/8}.$$

The inner sum on g is a geometric progression with sum 8 if $8|sd^2 - n$, and 0 if $8 \nmid sd^2 - n$. Since d is odd, $d^2 \equiv 1 \pmod 8$. Thus the inner sum is 8 if $n \equiv s \pmod 8$, and 0 if $n \not\equiv s \pmod 8$. Since also $\left(\frac{8j}{d}\right) = \left(\frac{4}{d}\right)\left(\frac{2j}{d}\right) = \left(\frac{2}{d}\right)^2\left(\frac{2j}{d}\right) = \left(\frac{2j}{d}\right)$, it follows that

$$D_d(n) = \begin{cases} d^{-s/2}e^{-\pi ids/4} \displaystyle\sum_{\substack{0 \leqslant j < d \\ (j,d)=1}} \left(\frac{2j}{d}\right)^s e^{-2\pi inj/d} & \text{if } n \equiv s \pmod 8, \\ 0 & \text{if } n \not\equiv s \pmod 8. \end{cases}$$

To complete the proof it suffices to show that $D_d(n) = e^{-\pi is/2}B_d(n)$ when $n \equiv s \pmod 8$. $B_d(n)$ was defined by

$$B_d(n) = d^{-s/2} \sum_{0 \leqslant j < 2d}^{*} \bar{v}_s(M_{d,j})e^{2\pi inj/d},$$

where in the summation j and d are of opposite parity and $(j,d) = 1$. Since d is odd, j is even and we have

$$\bar{v}_s(M_{d,j}) = \left(\frac{j}{d}\right)^s e^{\pi ids/4}$$

Thus

$$\begin{aligned} B_d(n) &= d^{-s/2} e^{\pi i d s/4} \sum_{\substack{0 \leqslant j < d \\ (j,d)=1}} \left(\frac{2j}{d}\right)^s e^{2\pi i n j/d} \\ &= d^{-s/2} e^{\pi i d s/4} \sum_{\substack{0 \leqslant j < d \\ (j,d)=1}} \left(\frac{-2j}{d}\right)^s e^{-2\pi i n j/d} \\ &= d^{-s/2} (-1)^{\frac{d-1}{2}s} e^{\pi i d s/4} \sum_{\substack{0 \leqslant j < d \\ (j,d)=1}} \left(\frac{2j}{d}\right)^s e^{-2\pi i n j/d} \\ &= e^{\pi i s/2} D_d(n) \qquad \text{if } n \equiv s \pmod 8. \end{aligned}$$

This completes the proof.

Theorem 17. For $s \geqslant 9$, $\psi_s(\tau) \neq \vartheta^s(\tau)$.

Proof. We begin with expansion (17) of $\psi_s(\tau)$ at the cusp -1. Since $8\{s/8\}$ is the least nonnegative residue of s modulo 8 we can write each positive integer $n \equiv s \pmod 8$ in the form $n = 8m + 8\{s/8\}$, with $m + \{s/8\} > 0$. Thus (17) becomes

$$\psi_s(\tau) = (\tau + 1)^{-s/2} \sum_{m+\{s/8\}>0} (-1)^{-s/2} \varphi_s\left(8m + 8\left\{\frac{s}{8}\right\}\right) \times \exp\left[2\pi i\left(m + \left\{\frac{s}{8}\right\}\right)\left(\frac{-1}{\tau+1}\right)\right]. \tag{23}$$

This incidentally is simply the expansion (18) rewritten in a more explicit form. That is, we have shown that

$$a_s(m) = (-1)^{-s/2} \varphi_s\left(8m + 8\left\{\frac{s}{8}\right\}\right),$$

where $a_s(m)$ is the coefficient that occurs in (18). Using the definition of $\varphi_s(n)$ given in Theorem 11 and the result of Lemma 16, we find that

$$\begin{aligned} a_s(m) &= (-1)^{-s/2} \frac{(2\pi)^{s/2}}{\Gamma(s/2)} \left(m + \left\{\frac{s}{8}\right\}\right)^{s/2-1} \sum_{\substack{d>0 \\ d \text{ odd}}} e^{-\pi i s/2} B_d\left(8m + 8\left\{\frac{s}{8}\right\}\right) \\ &= \frac{(2\pi)^{s/2}}{\Gamma(s/2)} \left(m + \left\{\frac{s}{8}\right\}\right)^{s/2-1} \sum_{\substack{d>0 \\ d \text{ odd}}} B_d\left(8m + 8\left\{\frac{s}{8}\right\}\right), \end{aligned} \tag{24}$$

since, by our branch convention, $(-1)^{-s/2} = e^{\pi i s/2}$.

Suppose $s \geqslant 6$. Then, as in the proof of Theorem 14, we have

$$|a_s(m)| \geqslant \frac{(2\pi)^{s/2}}{\Gamma(s/2)}\left(m + \left\{\frac{s}{8}\right\}\right)^{s/2-1}\left[\left|B_1\left(8m + 8\left\{\frac{s}{8}\right\}\right)\right| - \sum_{\substack{d \geqslant 2 \\ d \text{ odd}}}\left|B_d\left(8m + 8\left\{\frac{s}{8}\right\}\right)\right|\right]$$

$$> \frac{(2\pi)^{s/2}}{\Gamma(s/2)}\left(m + \left\{\frac{s}{8}\right\}\right)^{s/2-1}\left[2 - \zeta\left(\frac{s}{2} - 1\right)\right]$$

$$> \frac{1}{6}\frac{(2\pi)^{s/2}}{\Gamma(s/2)}\left(m + \left\{\frac{s}{8}\right\}\right)^{s/2-1} > 0,$$

since $m + \{s/8\} > 0$. In particular this shows that the *first* coefficient occurring in (23) [$a_s(1)$ if $s \equiv 0 \pmod 8$ and $a_s(0)$ if $s \not\equiv 0 \pmod 8$] is not equal to zero. Thus the expansion (23) of $\psi_s(\tau)$ begins with a term of the form $\exp\{2\pi i[-1/(\tau + 1)]\}$ if $s \equiv 0 \pmod 8$ and a term of the form $\exp\{2\pi i\{s/8\} \times[-1/(\tau + 1)]\}$ if $s \not\equiv 0 \pmod 8$. On the other hand, the expansion of $\vartheta^s(\tau)$ at -1 begins with a term of the form $\exp\{2\pi i(s/8)[-1/(\tau + 1)]\}$, as we see immediately from (20). If $s \geqslant 9$, then of course $s/8 > 1 > \{s/8\}$, so that $\vartheta^s(\tau)$ has a *strictly greater* order of vanishing at -1 than does $\psi_s(\tau)$. Thus for $s \geqslant 9$, $\psi_s(\tau) \neq \vartheta^s(\tau)$, and the proof is complete.

In the proof of Theorem 17 we have also derived

Corollary 18. At the cusp -1, $\psi_s(\tau)$ has the expansion

$$\psi_s(\tau) = (\tau + 1)^{-s/2} \sum_{m+\{s/8\}>0} a_s(m) \exp\left[2\pi i\left(m + \left\{\frac{s}{8}\right\}\right)\left(\frac{-1}{\tau + 1}\right)\right],$$

with $a_s(m)$ given by (24). Furthermore, if $s \geqslant 6$, then $a_s(m) \neq 0$ for all values of m.

As a final result concerning $\psi_s(\tau)$ we have

Corollary 19. For $5 \leqslant s \leqslant 8$, $\psi_s(\tau) \neq 0$ in $\mathscr{H}$; however, if $s \geqslant 9$, then $\psi_s(\tau)$ has a zero in $\mathscr{H}$.

Proof. The first statement follows from Theorem 3 and the fact that $\vartheta(\tau) \neq 0$ in $\mathscr{H}$.

To prove the second statement suppose $s \geqslant 9$ and $\psi_s(\tau) \neq 0$ in $\mathscr{H}$. Then $\vartheta^s(\tau)/\psi_s(\tau)$ is a modular function on Γ_ϑ, regular in $\mathscr{H}$, and with Fourier expansions at ∞ and -1 which contain no terms with negative exponents. Then by Theorem 7 of Chapter 2 it follows that $\vartheta^s(\tau)/\psi_s(\tau)$ is a constant; in fact, then $\vartheta^s(\tau)/\psi_s(\tau) = 1$. This contradicts Theorem 17. Thus for $s \geqslant 9$ $\psi_s(\tau)$ has a zero in $\mathscr{H}$.

Chapter 6

THE ORDER OF MAGNITUDE OF $p(n)$

1. A SIMPLE INEQUALITY FOR $p(n)$

The number-theoretic function $p(n)$ has already been defined in Section 1 of Chapter 3. By Corollary 2 of Chapter 3 the modular form $\eta^{-1}(\tau)$ serves as a generating function for $p(n)$ in the sense that

$$\eta^{-1}(\tau) = \sum_{m=-1}^{\infty} p(m+1)e^{2\pi i(m+23/24)\tau}, \qquad \tau \in \mathscr{H}.$$

Another way of writing this is

$$\prod_{m=1}^{\infty} (1 - x^m)^{-1} = \sum_{n=0}^{\infty} p(n)x^n, \qquad |x| < 1. \tag{1}$$

This occurrence of $p(n)$ as the coefficient in the expansion of a modular form is, of course, the reason that the theory of modular forms, as developed in Chapter 2, can be applied to yield information about $p(n)$.

In Chapter 5 we obtained identities and asymptotic estimates for $r_s(n)$, with $s \geqslant 5$. In this chapter we obtain an excellent asymptotic estimate for $p(n)$ by means of the "circle method." This method has been improved by Rademacher [*Proc. London Math Soc.*, *43* (1937), pp. 241–254] to give an *exact formula* for $p(n)$, but we do not give a proof, since Rademacher's "improved circle method" is rather complicated, and since excellent expositions are already available [e.g., J. Lehner, *Discontinuous Groups and Automorphic Functions* (Providence: American Mathematical Society, 1964), pp. 302–313 and 350-351].

The exact formula is

(2) $$p(n) = \frac{e^{\pi i/4}}{2} \sum_{k=1}^{\infty} A_k(n) k^{1/2} \frac{d}{dn}\left\{\frac{\sinh(\pi\sqrt{\frac{2}{3}}\sqrt{n - \frac{1}{24}}/k)}{\sqrt{n - \frac{1}{24}}}\right\},$$

for all n, where

$$A_k(n) = \sum_{\substack{h(\mathrm{mod}\, k) \\ (h,k)=1}} v_\eta(M_{k,-h}) \exp\left\{\frac{-(n - \frac{1}{24})h - \frac{1}{24}h'}{k}\right\}.$$

Here $M_{k,-h} = \begin{pmatrix} h' & * \\ k & -h \end{pmatrix} \in \Gamma(1)$, and v_η is the multiplier system for $\eta(\tau)$. An explicit formula for v_η is contained in Theorem 2 of Chapter 4.

Before we state and prove the promised asymptotic formula for $p(n)$, we prove the much simpler

Proposition 1. For all integers $n \geqslant 1$ we have

$$p(n) < e^{\sqrt{2/3}\pi\sqrt{n}}.$$

Proof. Denoting the left-hand side of (1) by $G(x)$, we have, for $|x| < 1$,

$$\log G(x) = -\sum_{m=1}^{\infty} \log(1 - x^m) = \sum_{m=1}^{\infty}\sum_{j=1}^{\infty} \frac{x^{mj}}{j}$$
$$= \sum_{j=1}^{\infty} \frac{1}{j} \sum_{m=1}^{\infty} x^{mj} = \sum_{j=1}^{\infty} \frac{1}{j}\left(\frac{x^j}{1 - x^j}\right),$$

where we have used the absolute convergence of the double sum to interchange the order of summation.

We now restrict the range of x to $0 < x < 1$. In this range we claim that

(3) $$jx^{j-1}(1 - x) < 1 - x^j, \qquad \text{for } j \geqslant 1.$$

To prove (3) simply apply the mean value theorem to the function t^j for t in the interval $[x,1]$. This tells us that there exists t such that $x < t < 1$ and

$$\frac{1 - x^j}{1 - x} = jt^{j-1}.$$

Since $t > x$, we have that $jt^{j-1} > jx^{j-1}$ and (3) follows. As a consequence of (3) we conclude that

$$\frac{1}{j}\left(\frac{x^j}{1 - x^j}\right) < \frac{x}{j^2(1 - x)},$$

so that, for $0 < x < 1$,

$$\log G(x) < \sum_{j=1}^{\infty} \frac{x}{j^2(1 - x)} = \left(\frac{x}{1 - x}\right) \sum_{j=1}^{\infty} \frac{1}{j^2} = \frac{\pi^2}{6}\left(\frac{x}{1 - x}\right).$$

Now it is obvious from (1) that for $0 < x < 1$, $p(n) \leqslant G(x)/x^n$. Hence $\log p(n) \leqslant \log G(x) - n \log x$, for $0 < x < 1$. Applying the mean value theorem to the function $\log t$ for t in the interval $[1,1/x]$, we obtain

$$-\log x = \log \frac{1}{x} < \frac{1}{x} - 1 = \frac{1-x}{x},$$

for $0 < x < 1$. Thus

$$\log p(n) < \frac{\pi^2}{6}\left(\frac{x}{1-x}\right) + n\left(\frac{1-x}{x}\right). \tag{4}$$

We now make the special choice of x for which the two terms on the right-hand side of (4) become equal. That is, we choose

$$x = \frac{\sqrt{6n}}{\pi + \sqrt{6n}} < 1,$$

and (4) becomes

$$\log p(n) < \sqrt{\tfrac{2}{3}}\pi\sqrt{n}.$$

Thus $p(n) < e^{\sqrt{2/3}\pi\sqrt{n}}$, and the proof is complete.

2. THE ASYMPTOTIC FORMULA FOR $p(n)$

A study of the infinite series on the right-hand side of formula (2) would indicate that the term corresponding to $k = 1$ is, by itself, a good estimate of $p(n)$. A straightforward calculation, using the fact that $A_1(n) = v_\eta(T) = e^{-\pi i/4}$, shows that this term is

$$\frac{\cosh(\pi v\sqrt{n - \frac{1}{24}})}{2\sqrt{3}(n - \frac{1}{24})} - \frac{\sinh(\pi v\sqrt{n - \frac{1}{24}})}{2\sqrt{2}\pi(n - \frac{1}{24})^{3/2}}, \tag{5}$$

where $v = \sqrt{\frac{2}{3}}$. As a matter of convenience, v will continue to have this meaning for the remainder of this chapter. For large n the expression (5) is very close to

$$\frac{e^{\pi v\sqrt{n-(1/24)}}}{4\sqrt{3}(n - \frac{1}{24})} - \frac{e^{\pi v\sqrt{n-(1/24)}}}{4\sqrt{2}\pi(n - \frac{1}{24})^{3/2}} = \frac{1}{4\sqrt{3}}\,\frac{e^{\pi v\sqrt{n-(1/24)}}}{(n - \frac{1}{24})}\left\{1 - \frac{1}{\pi v\sqrt{n - \frac{1}{24}}}\right\},$$

and in fact this is the main term of our asymptotic formula, which we state as

Theorem 2. As $n \to +\infty$, we have

$$p(n) = \frac{1}{4\sqrt{3}}\,\frac{e^{\pi v\sqrt{n-(1/24)}}}{(n - \frac{1}{24})}\left\{1 - \frac{1}{\pi v\sqrt{n - \frac{1}{24}}}\right\} + O(e^{(\pi v/2)\sqrt{n-(1/24)}}). \tag{6}$$

Remark. The formula (6) means that there exists a constant K, independent of n, such that for all $n \geqslant 0$,

$$|p(n) - [\text{first term on the right-hand side of (6)}]| < Ke^{(\pi\nu/2)\sqrt{n-(1/24)}}.$$

As an immediate consequence of Theorem 2 we obtain

Corollary 3. (a) As $n \to +\infty$, we have

$$p(n) = \frac{1}{4\sqrt{3}} \frac{e^{\pi\nu\sqrt{n-(1/24)}}}{(n - \frac{1}{24})} + O\left(\frac{e^{\pi\nu\sqrt{n}}}{n^{3/2}}\right).$$

(b) As $n \to +\infty$, $p(n) \sim \dfrac{1}{4\sqrt{3}n} e^{\pi\nu\sqrt{n}}$; that is,

$$\lim_{n\to+\infty} \frac{n4\sqrt{3}p(n)}{e^{\pi\nu\sqrt{n}}} = 1.$$

In the proof of Theorem 2 we need

Lemma 4. Let α and β be fixed positive real numbers. With $\rho > 0$, let C_ρ be the open contour $L_\rho^+ \cup C_\rho' \cup L_\rho^-$ in the plane, traversed in the counterclockwise direction, where L_ρ^+ is the real axis from 0 to ρ, with argument $= 0$; C_ρ' is the circle of radius ρ centered at 0; and L_ρ^- is the real axis from ρ to 0, with argument $= 2\pi$. Then

$$\int_{C_\rho} e^{-\alpha u - \beta/u}\sqrt{u}\, du = \frac{\sqrt{\pi}}{\alpha}\left\{2\sqrt{\beta}\cosh(2\sqrt{\alpha\beta}) - \frac{1}{\sqrt{\alpha}}\sinh(2\sqrt{\alpha\beta})\right\}$$

(see Figure 5).

Remark. The right-hand side is independent of ρ. The left-hand side is also independent of ρ, by Cauchy's theorem, since the only singularity of the integrand is the point 0.

Proof. Consider the integral

$$I_0 = -\int_{C_\rho} e^{-\alpha u - \beta/u}(1/\sqrt{u})\, du,$$

as a function of α. L'Hôpital's rule shows that the integrand in I_0 approaches 0 as $u \to 0+$. Thus the integral I_0 exists. We want to calculate $dI_0/d\alpha$, where, temporarily, α is thought of as a variable with positive real values. By definition,

$$\frac{dI_0}{d\alpha} = -\lim_{\varepsilon\to 0}\int_{C_\rho} e^{-\alpha u - \beta/u}\frac{1}{\sqrt{u}}\left(\frac{e^{-\varepsilon u} - 1}{\varepsilon}\right) du.$$

By the bounded convergence theorem [E. C. Titchmarsh, *The Theory of Functions* (2nd ed., rev.; New York: Oxford University Press, 1952),

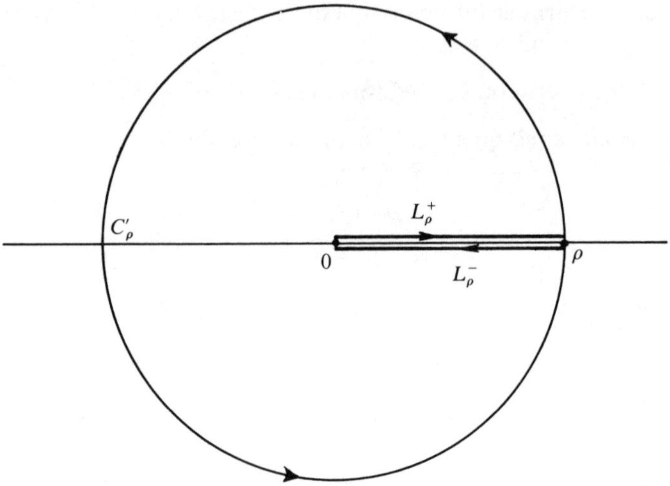

Figure 5. The contour C_ρ.

pp. 337–338], or by the Lebesgue dominated convergence theorem, we may take the limit on ε inside the integral sign to obtain

$$\frac{dI_0}{d\alpha} = \int_{C_\rho} e^{-\alpha u - \beta/u} \sqrt{u}\, du,$$

the integral we want to calculate.

To help calculate I_0, we use the relations

$\alpha u + \beta/u = (\sqrt{\alpha u} + \sqrt{\beta/u})^2 - 2\sqrt{\alpha\beta} = (\sqrt{\alpha u} - \sqrt{\beta/u})^2 + 2\sqrt{\alpha\beta}$, and write $I_0 = -2\alpha^{-1/2} \int_{C_\rho} e^{-\alpha u - \beta/u}\, d(\sqrt{\alpha u}) = \alpha^{-1/2}(e^{-2\sqrt{\alpha\beta}} I_1 - e^{2\sqrt{\alpha\beta}} I_2)$, where

$$I_1 = \int_{C_\rho} e^{-(\sqrt{\alpha u} - \sqrt{\beta/u})^2} d(\sqrt{\beta/u} - \sqrt{\alpha u}),$$

$$I_2 = \int_{C_\rho} e^{-(\sqrt{\alpha u} + \sqrt{\beta/u})^2} d(\sqrt{\alpha u} + \sqrt{\beta/u}).$$

By Cauchy's theorem both I_1 and I_2 are independent of ρ. Thus we may choose $\rho > 0$ for our convenience in calculating each integral. In the case of I_1 choose $\rho = \beta/\alpha$, and put $v = \sqrt{\beta/u} - \sqrt{\alpha u}$. On L_ρ^+, as u goes from 0 to ρ, v traverses the real axis from $+\infty$ to $\sqrt{\alpha} - \sqrt{\beta}$. As u traverses C_ρ in the counterclockwise direction, v goes from $\sqrt{\alpha} - \sqrt{\beta}$ to $\sqrt{\beta} - \sqrt{\alpha}$

along an ellipse with center at 0 and through the point $-i(\sqrt{\alpha}+\sqrt{\beta})$. As u goes from ρ to 0 on L_ρ^-, v goes from $\sqrt{\beta}-\sqrt{\alpha}$ to $-\infty$ along the real axis. Thus, using Cauchy's theorem to replace the elliptical part of the path by a straight-line segment, we obtain

$$I_1 = \int_{+\infty}^{-\infty} e^{-v^2}\,dv = -\int_{-\infty}^{\infty} e^{-v^2}\,dv = -\sqrt{\pi}.$$

To calculate I_2 we choose $\rho = 1$ and put $v = \sqrt{\alpha u} + \sqrt{\beta/u}$. Calculations similar to those used for I_1 show that, also,

$$I_2 = \int_{+\infty}^{-\infty} e^{-v^2}\,dv = -\sqrt{\pi}.$$

Hence

$$I_0 = \frac{\sqrt{\pi}}{\sqrt{\alpha}}(e^{2\sqrt{\alpha\beta}} - e^{-2\sqrt{\alpha\beta}}) = \frac{2\sqrt{\pi}}{\sqrt{\alpha}}\sinh(2\sqrt{\alpha\beta}).$$

Differentiating this equation with respect to α, we find that

$$\int_{C_\rho} e^{-\alpha u-\beta/u}\sqrt{u}\,du = \frac{\sqrt{\pi}}{\alpha}\left\{2\sqrt{\beta}\cosh(2\sqrt{\alpha\beta}) - \frac{1}{\sqrt{\alpha}}\sinh(2\sqrt{\alpha\beta})\right\},$$

and the proof is complete.

3. PROOF OF THEOREM 2

By (1) and the Cauchy integral formula,

$$p(n) = \frac{1}{2\pi i}\int_C \frac{G(t)}{t^{n+1}}\,dt, \qquad \text{for } n \geqslant 0,$$

where C is any circle centered at 0, of radius <1. We make the change of variable $t = e^{2\pi i z}$, and this becomes

$$p(n) = \int_L e^{-2\pi i n z} G(e^{2\pi i z})\,dz = \int_L e^{\pi i z/12} e^{-2\pi i n z}\frac{1}{\eta(z)}\,dz = \int_L e^{-2\pi i m z}\frac{1}{\eta(z)}\,dz,$$

where $m = n - \frac{1}{24}$ and L is a straight-line segment in $\mathscr{H}$ parallel to the real axis, and going from $-\frac{1}{2} + i\varepsilon$ to $\frac{1}{2} + i\varepsilon$, $\varepsilon > 0$. We next break L into three parts, $L = L_1 \cup L_2 \cup L_3$, where L_1 goes from $-\frac{1}{2} + i\varepsilon$ to $-\sqrt{2\varepsilon} + i\varepsilon$, L_2 goes from $-\sqrt{2\varepsilon} + i\varepsilon$ to $\sqrt{2\varepsilon} + i\varepsilon$, and L_3 goes from $\sqrt{2\varepsilon} + i\varepsilon$ to $\frac{1}{2} + i\varepsilon$. We choose $0 < \varepsilon < \frac{1}{8}$, so that $\sqrt{2\varepsilon} < \frac{1}{2}$. The proof is divided into several steps. We assume, as we may, that $n \geqslant 1$.

Step 1. Here we show that $\int_{L_1 \cup L_3} e^{-2\pi imz}(1/\eta(z))\,dz$ is small enough to be included in the O-term. Let $z \in L_1 \cup L_3$; since $\mathscr{R}(\Gamma(1))$ is a F.R. for $\Gamma(1)$, there exists $M = \begin{pmatrix} a & b \\ c & d \end{pmatrix} \in \Gamma(1)$ such that $z' = Mz \in \overline{\mathscr{R}(\Gamma(1))}$. With $z' = x' + iy'$, this implies in particular, that $y' \geqslant \sqrt{\frac{3}{2}}$, $|x'| \leqslant \frac{1}{2}$. Since $y = \operatorname{Im} z = \varepsilon < \frac{1}{8}$, we have also that $y' > y$. On the other hand, $y' = y|cz + d|^{-2}$, so that $|cz + d|^2 = y/y' < 1$. Hence $|cz + d|^{1/2} < 1$. We apply the transformation formula for $\eta(z)$,

$$\eta(Mz) = v_\eta(M)(cz + d)^{1/2}\eta(z),$$

to conclude that

$$|1/\eta(z)| = |v_\eta(M)||cz + d|^{1/2}|1/\eta(z')| < |1/\eta(z')|.$$

But also, from the power-series expression for $\eta(z)$,

$$|\eta(z')|^{-1} = |e^{-\pi iz'/12}|\left|\sum_{n=0}^{\infty} p(n)e^{2\pi inz'}\right| \leqslant e^{\pi y'/12}\sum_{n=0}^{\infty} p(n)e^{-2\pi ny'}$$

$$\leqslant e^{\pi y'/12}\sum_{n=0}^{\infty} p(n)e^{-\pi n\sqrt{3}} = K_1 e^{\pi y'/12},$$

with K_1 a positive constant. Thus, for $z \in L_1 \cup L_3$,

$$|\eta(z)|^{-1} < |\eta(z')|^{-1} \leqslant K_1 e^{\pi y'/12},$$

so that

$$\left|\int_{L_1 \cup L_3} e^{-2\pi imz}\frac{1}{\eta(z)}\,dz\right| \leqslant K_1 \int_{L_1 \cup L_3} e^{2\pi my} e^{\pi y'/12}\,dx$$

$$= K_1 e^{2\pi m\varepsilon}\int_{L_1 \cup L_3} e^{\pi y'/12}\,dx$$

$$\leqslant K_1 e^{2\pi m\varepsilon} \sup_{z \in L_1 \cup L_3}\{e^{\pi y'/12}\}.$$

We claim that with $z \in L_1 \cup L_3$, $|c| \geqslant 2$. Suppose $|c| < 2$, that is, $c = 0$ or $c = \pm 1$. If $c = 0$, then $d = \pm 1$, and $y' = y = \varepsilon < \frac{1}{8}$, contradicting the fact that $y' \geqslant \sqrt{\frac{3}{2}}$. If $c = \pm 1$ we consider the two possibilities $d = 0$ and $|d| \geqslant 1$. If $|d| \geqslant 1$, then $|cx + d| \geqslant |d| - |cx| \geqslant 1 - \frac{1}{2} = \frac{1}{2}$, so that

$$\frac{1}{y'} = \frac{(cx + d)^2}{y} + c^2 y \geqslant \frac{1}{4y} + y = \frac{1}{4\varepsilon} + \varepsilon \geqslant 1.$$

Hence $y' \leqslant 1$, a contradiction. If $c = \pm 1$ and $d = 0$, then since $|x| \geqslant \sqrt{2\varepsilon}$,

$$\frac{1}{y'} = \frac{x^2}{y} + y \geqslant \frac{2\varepsilon}{y} + y = 2 + \varepsilon > 2.$$

Hence $y' < \frac{1}{2}$, again contradicting the inequality $y' \geqslant \sqrt{\frac{3}{2}}$.

Thus for $z \in L_1 \cup L_3$, $|c| \geqslant 2$, so that

$\dfrac{1}{y'} = \dfrac{1}{y}\{(cx + d)^2 + c^2y^2\} \geqslant c^2 y \geqslant 4y = 4\varepsilon$, and consequently, $y' \leqslant 1/4\varepsilon$.

From this we conclude that

$$\left| \int_{L_1 \cup L_3} e^{-2\pi i m z} \frac{1}{\eta(z)} dz \right| \leqslant K_1 e^{\pi[2m\varepsilon + (1/48\varepsilon)]}.$$

We determine ε so as to minimize the exponent. The appropriate choice is $\varepsilon = 1/\sqrt{96m} < \frac{1}{8}$, and henceforth ε will have this value. With ε so chosen we obtain

$$\left| \int_{L_1 \cup L_3} e^{-2\pi i m z} \frac{1}{\eta(z)} dz \right| \leqslant K_1 e^{\pi v \sqrt{m}/2},$$

where, as before, $v = \sqrt{\frac{2}{3}}$.

Step 2. We next show that on L_2 we may replace the integrand by a far simpler function and still remain within the allowed error. Put

$$I = \int_{L_2} e^{-2\pi i m z} \frac{1}{\eta(z)} dz.$$

Using the transformation law of $\eta(z)$ under $T = \begin{pmatrix} 0 & -1 \\ 1 & 0 \end{pmatrix}$, we have $\eta(-1/z) = e^{-\pi i/4} z^{1/2} \eta(z)$, or

$$1/\eta(z) = \frac{e^{-\pi i/4} z^{1/2}}{\eta(-1/z)}.$$

Thus

$$\begin{aligned} I &= e^{-\pi i/4} \int_{L_2} e^{-2\pi i m z} \sqrt{z} \frac{1}{\eta(-1/z)} dz \\ &= e^{-\pi i/4} \int_{L_2} e^{\pi i/12z - 2\pi i m z} \sqrt{z} G(e^{-2\pi i/z})\, dz \\ &= e^{-\pi i/4} \int_{L_2} e^{-2\pi i g(z)} \sqrt{z} G(e^{-2\pi i/z})\, dz, \end{aligned}$$

where $g(z) = mz - 1/24z$. Consider the function

$$e^{-2\pi i g(z)} G(e^{-2\pi i/z}) = e^{-2\pi i g(z)} + e^{-2\pi i g(z)} \sum_{k=1}^{\infty} p(k) e^{(-2\pi i/z)k}.$$

Put $z_1 = -1/z = x_1 + iy_1$. Since $z \in L_2$, $x^2 \leqslant 2\varepsilon$, $y = \varepsilon$, so that

$$y_1 = y/|z|^2 = \frac{y}{x^2 + y^2} \geqslant \frac{1}{24\varepsilon} > \frac{1}{3}.$$

It follows that

$$\begin{aligned}\left| e^{-2\pi i g(z)} \sum_{k=1}^{\infty} p(k) e^{-2\pi i k/z} \right| &= \left| e^{-2\pi i m z} \sum_{k=1}^{\infty} p(k) e^{(-2\pi i/z)(k - 1/24)} \right| \\ &\leqslant e^{2\pi m y} \sum_{k=1}^{\infty} p(k) e^{-2\pi(k - 1/24)y_1} \\ &< e^{2\pi m y} \sum_{k=1}^{\infty} p(k) e^{(-2\pi/3)(k - 1/24)} = K_2 e^{2\pi m y} \\ &= K_2 e^{2\pi m \varepsilon} = K_2 e^{2\pi m/\sqrt{96m}} = K_2 e^{\pi \nu \sqrt{m}/4}.\end{aligned}$$

Hence

$$\begin{aligned}&\left| I - e^{-\pi i/4} \int_{L_2} e^{-2\pi i g(z)} \sqrt{z}\, dz \right| \\ &\quad = \left| \int_{L_2} e^{-2\pi i g(z)} \sqrt{z} \left(\sum_{k=1}^{\infty} p(k) e^{-2\pi i k/z} \right) dz \right| \\ &\quad < K_2 e^{\pi \nu \sqrt{m}/4} \int_{L_2} |\sqrt{z}|\, dz < K_2 e^{\pi \nu \sqrt{m}/4}.\end{aligned}$$

Combining this inequality with the results of step 1, we obtain

(7) $$p(n) = e^{-\pi i/4} \int_{L_2} e^{-2\pi i g(z)} \sqrt{z}\, dz + O(e^{\pi \nu \sqrt{m}/2}), \qquad \text{as } n \to +\infty.$$

Step 3. We proceed to replace the path L_2 in (7) by a more convenient one, by making use of Cauchy's theorem. We temporarily introduce the new convention $-\pi/2 \leqslant \arg z \leqslant \frac{3}{2}\pi$. Since this agrees with our previously established convention in $\mathscr{H}$ and since $L_2 \subset \mathscr{H}$, this change in determination of $\arg z$ has no effect on the value of the integral occurring on the right-hand side of (7). Recall that L_2 is the horizontal line segment from $-\sqrt{2\varepsilon} + i\varepsilon$ to $\sqrt{2\varepsilon} + i\varepsilon$, with $\varepsilon = (96\text{m})^{-1/2} = \{96(n - \frac{1}{24}\}^{-1/2} < \frac{1}{8}$. Let

$0 < \delta < \sqrt{2\varepsilon}$, and let L_4 be the vertical line segment from $-\sqrt{2\varepsilon}$ to $-\sqrt{2\varepsilon} + i\varepsilon$, let L_5 be the vertical line segment from $\sqrt{2\varepsilon} + i\varepsilon$ to $\sqrt{2\varepsilon}$, let L_6 be the horizontal line segment on the real axis from $-\delta$ to $-\sqrt{2\varepsilon}$, let L_7 be the horizontal line segment on the real axis from $\sqrt{2\varepsilon}$ to δ, let L_8 be the vertical line segment from $-\delta - i(1 - \delta^2)^{1/2}$ to $-\delta$, let L_9 be the vertical line segment from δ to $\delta - i(1 - \delta^2)^{1/2}$, and finally let $C(\delta)$ be the unit circle traversed in the counterclockwise direction from $\delta - i(1 - \delta^2)^{1/2}$ to $-\delta - i(1 - \delta^2)^{1/2}$. Then, since $\delta < \sqrt{2\varepsilon}$,

$$L_8 \cup L_6 \cup L_4 \cup L_2 \cup L_5 \cup L_7 \cup L_9 \cup C(\delta)$$

is a simple closed curve in the plane, excluding 0 from its interior (see Figure 6). Thus, by Cauchy's theorem,

$$\int_{L_2} e^{-2\pi i g(z)} \sqrt{z}\, dz = -\left(\sum_{j=4}^{9} \int_{L_j} e^{-2\pi i g(z)} \sqrt{z}\, dz + \int_{C(\delta)} e^{-2\pi i g(z)} \sqrt{z}\, dz \right).$$

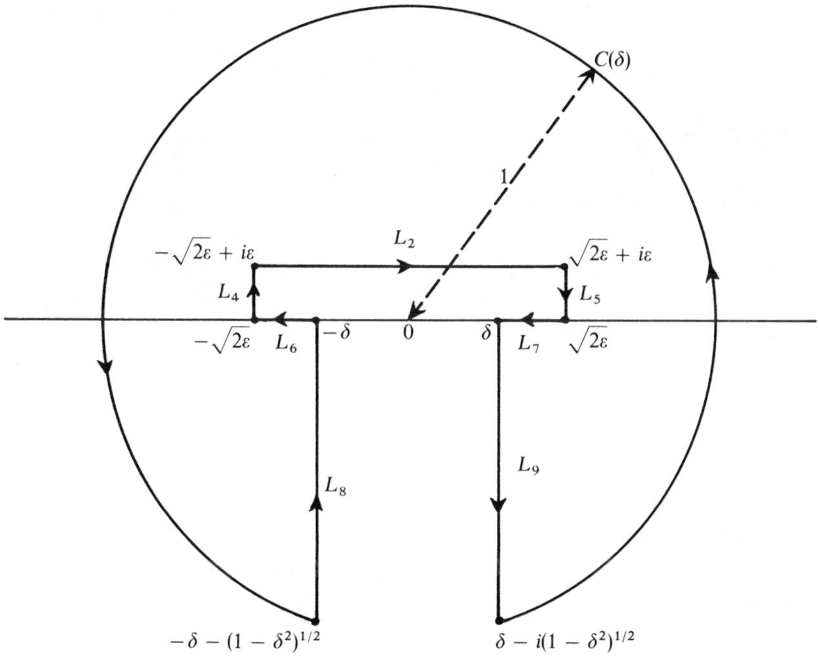

Figure 6

Now on $L_6 \cup L_7$, z is real, so that $g(z)$ is also real, and $|e^{-2\pi i g(z)}| = 1$. Thus

$$\left| \int_{L_6 \cup L_7} e^{-2\pi i g(z)} \sqrt{z}\, dz \right| \leqslant \int_{L_6 \cup L_7} |\sqrt{z}|\, dz \leqslant 1.$$

On L_4, $z = -\sqrt{2\varepsilon} + iy$, $0 \leqslant y \leqslant \varepsilon$, and on L_5, $z = \sqrt{2\varepsilon} + iy$, $0 \leqslant y \leqslant \varepsilon$. Thus, for $z \in L_4 \cup L_5$,

$$\operatorname{Im}(1/z) = \operatorname{Im}\left(\frac{\pm\sqrt{2\varepsilon} - iy}{2\varepsilon + y^2} \right) = \frac{-y}{2\varepsilon + y^2} \geqslant \frac{-\varepsilon}{2\varepsilon} = -\frac{1}{2},$$

so that, $\operatorname{Im}(g(z)) = \operatorname{Im}(mz - 1/24z) = m \operatorname{Im} z - \frac{1}{24} \operatorname{Im}(1/z) \leqslant m\varepsilon + \frac{1}{48}$. We conclude that, for $z \in L_4 \cup L_5$,

$|e^{-2\pi i g(z)}| = e^{2\pi \operatorname{Im}(g(z))} \leqslant e^{2\pi m\varepsilon} e^{\pi/24}$, and, consequently,

$$\left| \int_{L_4 \cup L_5} e^{-2\pi i g(z)} \sqrt{z}\, dz \right| \leqslant e^{\pi/24} e^{2\pi m\varepsilon} \int_{L_4 \cup L_5} |\sqrt{z}|\, dz = K_3 e^{2\pi m\varepsilon}$$

$$= K_3 e^{2\pi m/\sqrt{96m}} = K_3 e^{\pi\nu\sqrt{m}/4}.$$

We conclude from (7) that

$$p(n) = -e^{-\pi i/4} \int_{L_8 \cup L_9 \cup C(\delta)} e^{-2\pi i g(z)} \sqrt{z}\, dz + O(e^{\pi\nu\sqrt{m}/2}). \tag{8}$$

Step 4. This final step in the proof is the evaluation of the integral occurring in (8), with the help of Lemma 4. Put $u = iz$. Since $-\pi/2 \leqslant \arg z \leqslant \frac{3}{2}\pi$, we have $0 \leqslant \arg u \leqslant 2\pi$. With this substitution the integrand $-e^{-\pi/4} e^{-2\pi i g(z)} \sqrt{z}\, dz$ becomes $e^{-2\pi m u - \pi/12u} \sqrt{u}\, du$. The new path of integration is $C_1(\delta) = L_1^+(\delta) \cup C_1'(\delta) \cup L_1^-(\delta)$, where $L_1^+(\delta)$ is the horizontal line segment from $i\delta$ to $(1 - \delta^2)^{1/2} + i\delta$, $C_1'(\delta)$ is the unit circle traversed in the counterclockwise direction from $(1 - \delta^2)^{1/2} + i\delta$ to $(1 - \delta^2)^{1/2} - i\delta$, and $L_1^-(\delta)$ is the horizontal line segment from $(1 - \delta^2)^{1/2} - i\delta$ to $-i\delta$. Thus (8) becomes

$$p(n) = \int_{C_1(\delta)} e^{-2\pi m u - \pi/12u} \sqrt{u}\, du + O(e^{\pi\nu\sqrt{m}/2}). \tag{9}$$

We want to replace the path $C_1(\delta)$ in (9) by the path C_1 of Lemma 4 (that is, C_ρ, with $\rho = 1$). Since $p(n)$ and the O-term are independent of δ, we have, from (9),

$$p(n) = \lim_{\delta \to 0+} \int_{C_1(\delta)} e^{-2\pi m u - \pi/12u} \sqrt{u}\, du + O(e^{\pi\nu\sqrt{m}/2}).$$

We claim that

$$(10)\qquad \lim_{\delta\to 0+}\int_{C_1(\delta)} e^{-2\pi mu-\pi/12u}\sqrt{u}\,du = \int_{C_1} e^{-2\pi mu-\pi/12u}\sqrt{u}\,du.$$

If we can prove (10), then by Lemma 4,

$$\begin{aligned} p(n) &= \frac{\sqrt{\pi}}{2\pi m}\left\{2\sqrt{\pi/12}\cosh(\pi\nu\sqrt{m}) - \frac{1}{\sqrt{2\pi m}}\sinh(\pi\nu\sqrt{m})\right\} + O(e^{\pi\nu\sqrt{m}/2}) \\ &= \frac{1}{4\sqrt{3}m}e^{\pi\nu\sqrt{m}} - \frac{1}{4\pi\sqrt{2}m^{3/2}}e^{\pi\nu\sqrt{m}} + O(e^{\pi\nu\sqrt{m}/2}) \\ &= \frac{e^{\pi\nu\sqrt{m}}}{4\sqrt{3}m}\left\{1 - \frac{1}{\pi\nu m^{1/2}}\right\} + O(e^{\pi\nu\sqrt{m}/2}). \end{aligned}$$

Making the substitution $m = n - \frac{1}{24}$ produces formula (6), which was to be proved.

Thus we have reduced the proof of Theorem 2 to the proof of (10). In order to prove (10), consider first the integral along $L_1^+(\delta)$. We have

$$\begin{aligned} &\int_{L_1^+(\delta)} e^{-2\pi mu-\pi/12u}\sqrt{u}\,du \\ &\quad = (1-\delta^2)^{1/2}\int_{L_1^+} e^{-2\pi m(\sqrt{1-\delta^2}u+i\delta)}\exp\left[-\frac{\pi}{12(\sqrt{1-\delta^2}u+i\delta)}\right] \\ &\qquad \times\sqrt{(1-\delta^2)u+i\delta}\;du \\ &\quad \to \int_{L_1^+} e^{-2\pi mu-(\pi/12u)}\sqrt{u}\,du, \qquad \text{as } \delta\to 0_+, \end{aligned}$$

by the bounded convergence theorem. In the same way,

$$\lim_{\delta\to 0+}\int_{L_1^-(\delta)} e^{-2\pi mu-\pi/12u}\sqrt{u}\,du = \int_{L_1^-} e^{-2\pi mu-\pi/12u}\sqrt{u}\,du.$$

Finally, it is a trivial observation that

$$\lim_{\delta\to 0+}\int_{C_1'(\delta)} e^{-2\pi mu-\pi/12u}\sqrt{u}\,du = \int_{C_1'} e^{-2\pi mu-\pi/12u}\sqrt{u}\,du.$$

Thus (10) is proved and, with it, Theorem 2.

Chapter 7

THE RAMANUJAN CONGRUENCES FOR $p(n)$

1. STATEMENT OF THE CONGRUENCES

Ramanujan's original conjecture concerning the congruence properties of $p(n)$, as corrected by G. N. Watson [*J. Reine Angew. Math.*, *179* (1938), pp. 97–118], may be stated in the following way.

$$\text{(1)}\qquad \begin{aligned} &\text{(a) If } 24m \equiv 1 \pmod{5^n}, \text{ then } p(m) \equiv 0 \pmod{5^n}.\\ &\text{(b) If } 24m \equiv 1 \pmod{7^n}, \text{ then } p(m) \equiv 0 \pmod{7^{[(n+2)/2]}}.\\ &\text{(c) If } 24m \equiv 1 \pmod{11^n}, \text{ then } p(m) \equiv 0 \pmod{11^n}.\end{aligned}$$

As usual $[(n + 2)/2]$ denotes the largest integer $\leqslant (n + 2)/2$. In (1) m and n represent positive integers. In the article cited above Watson succeeded in proving (1a) and (1b) but left (1c) open. In fact, until quite recently (1c) was left unproved except for the cases $n = 1$ and $n = 2$. A. O. L. Atkin has now proved (1c) for all n [*Glasgow Math. J.*, *8* (1967), pp. 14–32], so that the proof of (1) has been completed.

In this chapter and the next one we present proofs of (1a) and (1b); we also give a proof of (1c), but *for the case* $n = 1$ *only*. Our proofs of (1a) and (1b) are taken from an unpublished manuscript of A. O. L. Atkin, entitled "On Watson's proof of properties of $p(n)$ modulo powers of 5 and 7, and some further results." It should be pointed out that although the Atkin manuscript is based upon the proofs of (1a) and (1b) given by Watson, it achieves a noteworthy simplification of Watson's proofs. Our proof of (1c) for the case $n = 1$ is one given by Morris Newman [*Can. J. Math.*, *9* (1957), pp. 68–70]. The general method of Newman (Sections 2 and 3 of this chapter), which leads to the proof of (1c) for $n = 1$, also is applied, together with

Theorem 7 of Chapter 2, to derive the "modular equations" for the primes 5 and 7. These modular equations are essential to the proofs of (1a) and (1b).

In this chapter we develop the modular function theory necessary in the proofs of (1a) and (1b). Also in Section 4 we carry out the proof of (1c) for the case $n = 1$ in its entirety. Chapter 8 will be devoted to the completion of the proofs of (1a) and (1b) and to the proofs of some related identities.

2. THE FUNCTIONS $\Phi_{p,r}(\tau)$ AND $h_p(\tau)$

Let p be a prime number greater than 3 and let $\Phi_{p,r}(\tau)$ be defined in $\mathscr{H}$ by

$$\Phi_{p,r}(\tau) = \{\eta(p\tau)/\eta(\tau)\}^r. \tag{2}$$

Theorem 1. If r is an integer such that $r(p-1) \equiv 0 \pmod{24}$, then $\Phi_{p,r}(\tau)$ is a modular form of degree 0 on the group $\Gamma_0(p)$. The multiplier system v of $\Phi_{p,r}$ is given by $v(V) = \left(\frac{d}{p}\right)^r$, for $V = \begin{pmatrix} a & b \\ c & d \end{pmatrix} \in \Gamma_0(p)$, where $\left(\frac{d}{p}\right)$ is Legendre's symbol. Furthermore, $\Phi_{p,r}$ is regular and nonzero in $\mathscr{H}$, has an expansion at the cusp ∞ beginning with the term $\exp\{\pi i(p-1)r\tau/12\}$, and has an expansion at the cusp 0 beginning with the term $p^{-r/2}\cdot \exp\{-\pi i(p-1)r(-1/\tau)/12p\}$.

Remarks. 1. Recall from Chapter 2, Example 2, that there is a S.F.R. of $\Gamma_0(p)$ in which ∞ and 0 are the *only* parabolic cusps.

2. If $r > 0$, then $\Phi_{p,r}$ has a zero at ∞ and a pole at 0. If $r < 0$, the situation is reversed.

3. If r is even, $\Phi_{p,r}$ is actually a modular *function* on $\Gamma_0(p)$.

Proof. Clearly $\Phi_{p,r}$ is regular and nonzero in $\mathscr{H}$, since $\eta(\tau)$ is regular and nonzero in $\mathscr{H}$. We first derive the transformation properties of $\Phi_{p,r}$. Let $V = \begin{pmatrix} a & b \\ c & d \end{pmatrix} \in \Gamma_0(p)$. If $c = 0$, then $\begin{pmatrix} a & b \\ c & d \end{pmatrix} = \begin{pmatrix} \pm 1 & b \\ 0 & \pm 1 \end{pmatrix}$, so that

$$\Phi_{p,r}(V\tau) = \Phi_{p,r}(\tau \pm b) = \{\eta(p\tau \pm pb)/\eta(\tau \pm b)\}^r$$

$$= \left\{\frac{e^{\pm \pi i p b/12}\eta(p\tau)}{e^{\pm \pi i b/12}\eta(\tau)}\right\}^r = e^{\pm b\pi i r(p-1)/12}\Phi_{p,r}(\tau)$$

$$= \Phi_{p,r}(\tau) = \left(\frac{1}{p}\right)^r \Phi_{p,r}(\tau),$$

since $r(p-1) \equiv 0 \pmod{24}$.

Suppose $c \neq 0$ and c is odd. Then by Theorem 2 of Chapter 4,

$$\eta(V\tau) = \left(\frac{d}{c}\right)^* \exp\left\{\frac{\pi i}{12}\left[(a + d)c - bd(c^2 - 1) - 3c\right]\right\}(c\tau + d)^{1/2}\eta(\tau).$$

Also $p(V\tau) = (pa\tau + pb)/(c\tau + d) = V_1(p\tau)$, where $V_1 = \begin{pmatrix} a & pb \\ c/p & d \end{pmatrix} \in \Gamma(1)$. Thus

$$\eta(pV\tau) = \eta(V_1(p\tau)) = \left(\frac{d}{c/p}\right)^* \exp\left\{\frac{\pi i}{12}\left[(a + d)\frac{c}{p} - pbd\left(\frac{c^2}{p^2} - 1\right) - \frac{3c}{p}\right]\right\}$$
$$\times (c\tau + d)^{1/2}\eta(p\tau),$$

and we conclude that

$$\Phi_{p,r}(V\tau) = \exp\left(\frac{\pi i}{12}E\right)\left\{\left(\frac{d}{c/p}\right)^*\left(\frac{d}{c}\right)^*\right\}^r \Phi_{p,r}(\tau),$$

where

$$E = r\left\{(a + d)\frac{c}{p} - (a + d)c + bd(c^2 - 1) - pbd\left(\frac{c^2}{p^2} - 1\right) + 3c - \frac{3c}{p}\right\}.$$

Since $p|c$ and $r(p - 1) \equiv 0 \pmod{24}$, it follows easily that $E \equiv 0 \pmod{24}$. In addition,

$$\left(\frac{d}{c}\right)^* = \left(\frac{d}{|c|}\right) = \left(\frac{d}{|c/p|}\right)\left(\frac{d}{p}\right) = \left(\frac{d}{c/p}\right)^*\left(\frac{d}{p}\right),$$

so that

$$\Phi_{p,r}(V\tau) = \left(\frac{d}{p}\right)^r \Phi_{p,r}(\tau).$$

Suppose $c \neq 0$ and c is even. Then

$$\eta(V\tau) = \left(\frac{c}{d}\right)_* \exp\left\{\frac{\pi i}{12}\left[(a + d)c - bd(c^2 - 1) + 3d - 3 - 3cd\right]\right\}$$
$$\times (c\tau + d)^{1/2}\eta(\tau),$$

and

$$\eta(pV\tau) = \eta(V_1(p\tau))$$
$$= \left(\frac{c/p}{d}\right)_* \exp\left\{\frac{\pi i}{12}\left[(a + d)\frac{c}{p} - pbd\left(\frac{c^2}{p^2} - 1\right) + 3d - 3 - \frac{3cd}{p}\right]\right\}$$
$$\times (c\tau + d)^{1/2}\eta(p\tau).$$

Thus

$$\Phi_{p,r}(V\tau) = \exp\left(\frac{\pi i}{12}E\right)\left\{\left(\frac{c/p}{d}\right)_*\left(\frac{c}{d}\right)_*\right\}^r \Phi_{p,r}(\tau),$$

where $E = r\left\{(a+d)c(1-1/p) - bd\left(\frac{c^2}{p} - p - c^2 + 1\right) + 3cd\left(1-\frac{1}{p}\right)\right\}$. As before $E \equiv 0 \pmod{24}$, and

$$\left(\frac{c}{d}\right)_* = \left(\frac{c}{|d|}\right)(-1)^{\frac{\operatorname{sign} c - 1}{2}\frac{\operatorname{sign} d - 1}{2}} = \left(\frac{c/p}{d}\right)_*\left(\frac{p}{|d|}\right).$$

Thus

$$\left(\frac{c/p}{d}\right)_*\left(\frac{c}{d}\right)_* = \left(\frac{p}{|d|}\right) = \left(\frac{d}{p}\right)(-1)^{\frac{p-1}{2}\frac{d-1}{2}},$$

and we conclude that

$$\Phi_{p,r}(V\tau) = \left(\frac{d}{p}\right)^r(-1)^{\frac{p-1}{2}\frac{d-1}{2}r}\Phi_{p,r}(\tau) = \left(\frac{d}{p}\right)^r\Phi_{p,r}(\tau),$$

since d is odd and $r(p-1)/2$ is even.

We now consider the expansions of $\Phi_{p,r}$ at the parabolic cusps ∞ and 0. To get the expansion at ∞ we simply insert the original definition of $\eta(\tau)$ into the expression (2) for $\Phi_{p,r}$. This gives

$$\begin{aligned}\Phi_{p,r}(\tau) &= \exp\left\{\frac{\pi i r}{12}(p-1)\tau\right\}\prod_{m=1}^{\infty}(1-e^{2\pi i m p\tau})^r(1-e^{2\pi i m\tau})^{-r} \\ &= \exp\left\{\frac{\pi i(p-1)r}{12}\tau\right\}\left(1+\sum_{n=1}^{\infty}a_n e^{2\pi i n\tau}\right)\end{aligned}$$

as the expansion of $\Phi_{p,r}$ at ∞. To obtain the expansion at 0 we consider $\Phi_{p,r}(-1/\tau)$ and use the transformation property of η under $T = \begin{pmatrix}0 & -1\\ 1 & 0\end{pmatrix}$. We have

$$\Phi_{p,r}(-1/\tau) = \left\{\eta\left(\frac{-1}{(\tau/p)}\right)\Big/\eta(-1/\tau)\right\}^r = p^{-r/2}\{\eta(\tau/p)/\eta(\tau)\}^r.$$

Inserting the product definition of $\eta(\tau)$, we find that

$$\begin{aligned}\Phi_{p,r}(-1/\tau) &= p^{-r/2}\exp\left\{\frac{\pi i(1-p)r}{12p}\tau\right\}\prod_{m=1}^{\infty}(1-e^{2\pi i m\tau/p})^r(1-e^{2\pi i m\tau})^{-r} \\ &= p^{-r/2}\exp\left\{\frac{-\pi i(p-1)r}{12p}\tau\right\}\left(1+\sum_{n=1}^{\infty}b_n e^{2\pi i n\tau/p}\right).\end{aligned}$$

We now replace τ by $-1/\tau$ to obtain the desired expansion at 0. This completes the proof.

With p a prime number greater than 3 we now define the function $h_p(\tau)$ by

$$h_p(\tau) = \eta(p^2\tau)/\eta(\tau). \tag{3}$$

Clearly $h_p(\tau)$ is regular and zero-free in $\mathscr{H}$. It is actually the case that $h_p(\tau)$ is a modular function on $\Gamma_0(p^2)$, but we here content ourselves with proving the invariance of $h_p(\tau)$ under transformations in $\Gamma_0(p^2)$, omitting a discussion of the behavior of $h_p(\tau)$ at the parabolic cusps of a S.F.R. of $\Gamma_0(p^2)$. The proof of the invariance of $h_p(\tau)$ under $\Gamma_0(p^2)$ is almost the same as the proof of the transformation properties of $\Phi_{p,r}(\tau)$.

Theorem 2. If $V \in \Gamma_0(p^2)$, then $h_p(V\tau) = h_p(\tau)$.

Proof. Let $V = \begin{pmatrix} a & b \\ c & d \end{pmatrix} \in \Gamma_0(p^2)$. As in the proof of Theorem 1 we utilize Theorem 2 of Chapter 4. In this case we write

$$p^2 V\tau = (p^2 a\tau + p^2 b)/(c\tau + d) = V_1(p^2\tau),$$

where $V_1 = \begin{pmatrix} a & p^2 b \\ c/p^2 & d \end{pmatrix} \in \Gamma(1)$. Thus $h_p(V\tau) = \dfrac{v_\eta(V_1)}{v_\eta(V)} h(\tau)$.

If c is odd, then c/p^2 is also odd, and Theorem 2 of Chapter 4 implies that

$$v_\eta(V) = \left(\frac{d}{c}\right)^* \exp\left\{\frac{\pi i}{12}\left[(a + d)c - bd(c^2 - 1) - 3c\right]\right\}$$

and

$$v_\eta(V_1) = \left(\frac{d}{c/p^2}\right)^* \exp\left\{\frac{\pi i}{12}\left[(a + d)\frac{c}{p^2} - p^2 bd\left(\frac{c^2}{p^4} - 1\right) - \frac{3c}{p^2}\right]\right\}.$$

Thus

$$\frac{v_\eta(V_1)}{v_\eta(V)} = \left(\frac{d}{c}\right)^* \left(\frac{d}{c/p^2}\right)^* \exp\left(\frac{\pi i}{12} E\right),$$

where $E = (a + d)c\left(\dfrac{1}{p^2} - 1\right) + bd\left(c^2 - 1 - \dfrac{c^2}{p^2} + p^2\right) + 3c\left(1 - \dfrac{1}{p^2}\right)$.

Since p is an odd number not divisible by 3, it follows that $p^2 \equiv 1 \pmod{24}$, and from this it follows immediately that $E \equiv 0 \pmod{24}$. On the other hand,

$$\left(\frac{d}{c}\right)^* = \left(\frac{d}{|c|}\right) = \left(\frac{d}{|c/p^2|}\right)\left(\frac{d}{p^2}\right) = \left(\frac{d}{c/p^2}\right)^*,$$

so that

$$\left(\frac{d}{c}\right)^*\left(\frac{d}{c/p^2}\right)^* = \left(\frac{d}{c/p^2}\right)^{*2} = 1.$$

Hence $h_p(V\tau) = h_p(\tau)$, if c is odd.

Suppose c is even. If $c = 0$, then $a = d = \pm 1$ and we have

$$h_p(V\tau) = h_p(\tau \pm b) = \exp\left\{\frac{\pi i}{12}(p^2 - 1)(\pm b)\right\} h_p(\tau) = h_p(\tau),$$

again because $p^2 \equiv 1 \pmod{24}$. If $c \neq 0$ we apply Theorem 2 of Chapter 4 once more to obtain

$$\upsilon_\eta(V) = \left(\frac{c}{d}\right)_* \exp\left\{\frac{\pi i}{12}\left[(a + d)c - bd(c^2 - 1) + 3d - 3 - 3cd\right]\right\}$$

and

$$\upsilon_\eta(V_1) = \left(\frac{c/p^2}{d}\right)_* \exp\left\{\frac{\pi i}{12}\left[(a + d)\frac{c}{p^2} - p^2 bd\left(\frac{c^2}{p^4} - 1\right) + 3d - 3 - \frac{3cd}{p^2}\right]\right\}.$$

Thus

$$\frac{\upsilon_\eta(V_1)}{\upsilon_\eta(V)} = \left(\frac{c/p^2}{d}\right)_* \left(\frac{c}{d}\right)_* \exp\left(\frac{\pi i}{12}E\right),$$

with $E = (a + d)c\left(\frac{1}{p^2} - 1\right) + bd\left(c^2 - 1 - \frac{c^2}{p^2} + p^2\right) + 3cd\left(1 - \frac{1}{p^2}\right)$.

As before $E \equiv 0 \pmod{24}$. Also,

$$\left(\frac{c}{d}\right)_* = \left(\frac{c}{|d|}\right)(-1)^{\frac{\operatorname{sign} c - 1}{2}\frac{\operatorname{sign} d - 1}{2}} = \left(\frac{p^2}{|d|}\right)\left(\frac{c/p^2}{d}\right)_* = \left(\frac{c/p^2}{d}\right)_*.$$

Therefore,

$$\left(\frac{c/p^2}{d}\right)_* \left(\frac{c}{d}\right)_* = \left(\frac{c/p^2}{d}\right)_*^2 = 1,$$

and we conclude that $h_p(V\tau) = h_p(\tau)$. The proof is complete.

3. THE FUNCTION $S_{p,r}(\tau)$

Let p be a prime number greater than 3 and let r be any integer. Let $W = \begin{pmatrix} 1 & 0 \\ 1 & 1 \end{pmatrix} \in \Gamma(1)$ and define the function $S_{p,r}(\tau)$ by means of

(4) $$S_{p,r}(\tau) = \sum_{k=0}^{p-1} h_p^r(W^{-pk}\tau), \qquad \tau \in \mathscr{H}.$$

For the sake of convenience we shall usually abbreviate $S_{p,r}(\tau)$ to $S(\tau)$. We have the following basic result concerning $S_{p,r}(\tau)$.

Theorem 3. $S_{p,r}(\tau)$ is a modular function with respect to $\Gamma_0(p)$ and regular in $\mathscr{H}$. If $r > 0$, $S_{p,r}(\tau)$ has at most a pole at 0 and is regular at ∞. If $r < 0$, $S_{p,r}(\tau)$ is regular at 0 and has at most a pole at ∞.

In the course of the proof of Theorem 3 we prove identities for $S_{p,r}(\tau)$ of independent interest and which we therefore state separately as

Theorem 4. Let the integer $p_r(n)$, $n \geqslant 0$, be defined by means of

$$\prod_{m=1}^{\infty} (1 - x^m)^r = \sum_{n=0}^{\infty} p_r(n)x^n, \tag{5}$$

and put $v = (p^2 - 1)/24$. Then if r is even,

$$S_{p,r}(\tau) = -p^{-r/2}e^{(\pi i r/4)(1-p)} + h_p^r(\tau) + p^{1-r/2}e^{(\pi i r/4)(1-p)} \prod_{n=1}^{\infty} (1 - x^n)^{-r} \sum_{\substack{n=0 \\ n \equiv rv (\mathrm{mod}\ p)}}^{\infty} p_r(n)x^n. \tag{6}$$

If r is odd,

$$S_{p,r}(\tau) = h_p^r(\tau) + e^{\pi i (p-1)^2/8} p^{(1-r)/2} e^{(\pi i r/4)(1-p)} \prod_{n=1}^{\infty} (1 - x^n)^{-r} \times \sum_{n=0}^{\infty} \left(\frac{rv - n}{p}\right) p_r(n)x^n. \tag{7}$$

Here $x = e^{2\pi i \tau}$.

Remark. In the notation of Theorem 4, $p(n) = p_{-1}(n)$.

The proof of Theorem 4 requires the following lemma, which is of an elementary nature. The proof is postponed until the end of this section.

Lemma 5. With k an integer define k' by

$$k' = \begin{cases} \text{least positive solution of } kx \equiv 1 \ (\mathrm{mod}\ p) & \text{if } (k,p) = 1, \\ 0 & \text{if } p \mid k. \end{cases}$$

(a) if k is odd, then

$$pk + k\left\{\frac{kk' - 1}{p}\right\} + k'(k^2 - 1)\left\{\frac{kk' - 1}{p}\right\} \equiv pk' \ (\mathrm{mod}\ 24).$$

(b) If k is even, put $k = k^* 2^\alpha$, where k^* is odd and α is a positive integer. Then

$$\left(\frac{k}{kk' - 1}\right) = (-1)^{\frac{(kk'-1)^2 - 1}{8}\alpha + \frac{k^* - 1}{2}\frac{kk'}{2}}.$$

(c) If k is even, then, with the notation of (b), we have

$$\frac{(kk'-1)^2-1}{2}3\alpha + 3(k^*-1)kk' - 2pk + 3 - 2k\left\{\frac{kk'-1}{p}\right\}$$
$$+ k'\left\{\frac{kk'+1}{p}\right\}(k^2-1) + 3\left\{\frac{kk'-1}{p}\right\} \equiv 3 - 3p + 3pk' \pmod{24}.$$

Note that the left-hand side of (b) is Jacobi's symbol, while in (a) and (c) braces have been used to avoid confusion with Jacobi's symbol.

Proof of Theorem 4. We write $h_p(\tau) = h(\tau)$. Note that

$$h(W^{-pk}\tau) = \frac{\eta(p^2W^{-pk}\tau)}{\eta(W^{-pk}\tau)},$$

from the definition of $h(\tau)$. Let k' be as in Lemma 5. Then for $(k,p) = 1$ we have

$$p^2(W^{-pk}\tau) = \frac{p^2\tau}{-pk\tau+1} = M_1(\tau - k'/p),$$

where $M_1 = \begin{pmatrix} -p & -k' \\ k & (kk'-1)/p \end{pmatrix} \in \Gamma(1)$. Thus for $1 \leqslant k \leqslant p-1$ we have

$$h(W^{-pk}\tau) = \frac{\eta(M_1(\tau - k'/p))}{\eta(W^{-pk}\tau)}.$$

Since W^{-pk}, $M_1 \in \Gamma(1)$, it follows from the fact that $\eta(\tau)$ is a modular form of degree $-\frac{1}{2}$, with multiplier system υ_η, on $\Gamma(1)$, that

$$\eta(M_1(\tau - k'/p)) = \upsilon_\eta(M_1)(k\tau - 1/p)^{1/2}\eta(\tau - k'/p)$$

and

$$\eta(W^{-pk}\tau) = \eta((-W^{pk})\tau) = \upsilon_\eta(-W^{-pk})(pk\tau - 1)^{1/2}\eta(\tau).$$

Thus for $1 \leqslant k \leqslant p-1$ we have

$$h(W^{-pk}\tau) = p^{-1/2}\frac{\upsilon_\eta(M_1)}{\upsilon_\eta(-W^{-pk})}\frac{\eta(\tau - k'/p)}{\eta(\tau)}. \tag{8}$$

We now apply Theorem 2 of Chapter 4 to reevaluate the right side of (8). We claim that, for $1 \leqslant k \leqslant p-1$,

$$\frac{\upsilon_\eta(M_1)}{\upsilon_\eta(-W^{-pk})} = \left(\frac{k'}{p}\right)e^{\pi i(1-p)/4}e^{\pi ipk'/12}. \tag{9}$$

Suppose first that k is odd. Then by Theorem 2 of Chapter 4,

$$v_\eta(M_1) = \left(\frac{(kk'-1)/p}{k}\right)^* \exp\left[\frac{\pi i}{12}\left\{\left(-p + \frac{kk'-1}{p}\right)k + k'\left(\frac{kk'-1}{p}\right) \times (k^2-1) - 3k\right\}\right],$$

and

$$v_\eta(-W^{-pk}) = \left(\frac{-1}{pk}\right)^* e^{(\pi i/12)(-2pk-3pk)} = (-1)^{\frac{pk-1}{2}} e^{-5\pi i pk/12}.$$

Now from Lemma 1 of Chapter 4 we have

$$\left(\frac{(kk'-1)/p}{k}\right)^* = \left(\frac{(kk'-1)/p}{k}\right) = (-1)^{\frac{k-1}{2}}\left(\frac{p}{k}\right)$$

$$= (-1)^{\frac{k-1}{2}}(-1)^{\frac{p-1}{2}\frac{k-1}{2}}\left(\frac{k}{p}\right) = (-1)^{\frac{k-1}{2}\frac{p+1}{2}}\left(\frac{k'}{p}\right).$$

Thus, after some rearrangement, we obtain

$$\frac{v_\eta(M_1)}{v_\eta(-W^{-pk})} = \left(\frac{k'}{p}\right) e^{\pi i(1-p)/4} e^{(\pi i/12)E},$$

where $E = pk + k\left\{\frac{kk'-1}{p}\right\} + k'(k^2-1)\left\{\frac{kk'-1}{p}\right\}$. By Lemma 5(a), $E \equiv pk' \pmod{24}$, and (9) is proved in the case when k is odd.

Suppose k is even. In this case,

$$v_\eta(M_1) = \left(\frac{k}{(kk'-1)/p}\right)_* \exp\left(\frac{\pi i}{12}\left\{\left(-p + \frac{kk'-1}{p}\right)k + k'\left[\frac{kk'-1}{p}\right](k^2-1) + 3\frac{kk'-1}{p} - 3 - 3k\frac{kk'-1}{p}\right\}\right),$$

and

$$v_\eta(-W^{-pk}) = \left(\frac{pk}{-1}\right)_* e^{\pi i(pk-6)/12}.$$

Now

$$\left(\frac{k}{(kk'-1)/p}\right)_* = \left(\frac{k}{(kk'-1)/p}\right) = \left(\frac{k}{p}\right)\left(\frac{k}{kk'-1}\right) = \left(\frac{k'}{p}\right)\left(\frac{k}{kk'-1}\right)$$

$$= \left(\frac{k'}{p}\right)(-1)^{\frac{(kk'-1)^2-1}{8}\alpha + \frac{k^*-1}{2}\frac{kk'}{2}},$$

by Lemma 5(b). Also, $\left(\frac{pk}{-1}\right)_* = \left(\frac{pk}{1}\right) = 1$. Thus, in this case,

$$\frac{v_\eta(M_1)}{v_\eta(-W^{-pk})} = \left(\frac{k'}{p}\right) e^{(\pi i/12)E},$$

with

$$E = \frac{(kk'-1)^2-1}{2}3\alpha + 3(k^*-1)kk' - 2pk + 3 - 2\frac{kk'-1}{p} + k'\left\{\frac{kk'-1}{p}\right\}(k^2-1) + 3\frac{kk'-1}{p}.$$

By Lemma 5(c), $E \equiv 3 - 3p + pk' \pmod{24}$, and (9) is proved in the case when k is odd.

Combining (8) and (9), we obtain the following interesting identity, for $1 \leqslant k \leqslant p-1$:

$$h(W^{-pk}\tau) = p^{-1/2} e^{(\pi i/4)(1-p)} e^{\pi i k' p/12} \left(\frac{k'}{p}\right) \frac{\eta(\tau - k'/p)}{\eta(\tau)}. \tag{10}$$

Thus

$$\begin{aligned} S(\tau) &= \sum_{k=0}^{p-1} h^r(W^{-pk}\tau) \\ &= h^r(\tau) + \sum_{k=1}^{p-1} p^{-r/2} e^{(\pi i/4)(1-p)r} e^{\pi i k' pr/12} \left(\frac{k'}{p}\right)^r \frac{\eta^r(\tau - k'/p)}{\eta^r(\tau)} \\ &= h^r(\tau) + p^{-r/2} e^{(\pi i/4)(1-p)r} \prod_{n=1}^{\infty} (1-x^n)^{-r} e^{-\pi i \tau r/12} \\ &\quad \times \sum_{k=1}^{p-1} e^{\pi i k' pr/12} e^{(\pi i/12)(\tau - k'/p)r} \left(\frac{k'}{p}\right)^r \sum_{n=0}^{\infty} p_r(n) x^n e^{-2\pi i k' n/p} \\ &= h^r(\tau) + p^{-r/2} e^{(\pi i/4)(1-p)r} \prod_{n=1}^{\infty} (1-x^n)^{-r} \\ &\quad \times \sum_{k=1}^{p-1} e^{\pi i k' pr/12} e^{-\pi i k' r/12p} \left(\frac{k'}{p}\right)^r \sum_{n=0}^{\infty} p_r(n) x^n e^{-2\pi i k' n/p}, \end{aligned}$$

or

$$\begin{aligned} S(\tau) &= h^r(\tau) + p^{-r/2} e^{(\pi i/4)(1-p)r} \prod_{n=1}^{\infty} (1-x^n)^{-r} \sum_{n=0}^{\infty} p_r(n) x^n \\ &\quad \times \sum_{k=1}^{p-1} \left(\frac{k'}{p}\right)^r \exp\left[\frac{2\pi i k' r}{p}\left(\frac{p^2-1}{24}\right)\right] e^{-2\pi i k' n/p}, \end{aligned} \tag{11}$$

where we have put $x = e^{2\pi i\tau}$ and used the definition of $\eta(\tau)$ as an infinite product and the definition (5) of $p_r(n)$.

Suppose first that r is even. Since $v = (p^2 - 1)/24$, the finite inner sum in (11) is

$$\sum_{k=1}^{p-1} e^{(2\pi ik'/p)(rv-n)} = \sum_{k=1}^{p-1} e^{(2\pi ik/p)(rv-n)} = \begin{cases} -1 & \text{if } p\nmid(rv-n), \\ p-1 & \text{if } p\mid(rv-n). \end{cases}$$

Thus for even r we obtain

$$S(\tau) = h^r(\tau) + p^{-r/2}e^{(\pi i/4)(1-p)r} \prod_{n=1}^{\infty}(1-x^n)^{-r}(p-1) \sum_{\substack{n=0 \\ p\mid(rv-n)}}^{\infty} p_r(n)x^n$$

$$- \prod_{n=1}^{\infty}(1-x^n)^{-r} p^{-r/2}e^{(\pi i/4)(1-p)r} \sum_{\substack{n=0 \\ p\nmid(rv-n)}}^{\infty} p_r(n)x^n$$

$$= h^r(\tau) + p^{1-r/2}e^{(\pi i/4)(1-p)r} \prod_{n=1}^{\infty}(1-x^n)^{-r} \sum_{\substack{n=0 \\ n\equiv rv(\text{mod } p)}}^{\infty} p_r(n)x^n$$

$$-p^{-r/2}e^{(\pi i/4)(1-p)r} \prod_{n=1}^{\infty}(1-x^n)^{-r} \sum_{n=0}^{\infty} p_r(n)x^n.$$

Since $\Pi_{n=1}^{\infty}(1-x^n)^{-r}\Sigma_{n=0}^{\infty} p_r(n)x^n = 1$, (6) follows and the proof of Theorem 4 is complete for the case when r is even.

Suppose r is odd. Then the inner sum in (11) is

$$\sum_{k=1}^{p-1}\left(\frac{k'}{p}\right)e^{(2\pi ik'/p)(rv-n)} = \sum_{k=1}^{p-1}\left(\frac{k}{p}\right)e^{(2\pi ik/p)(rv-n)}.$$

We now need to use the famous Gaussian sum formula, which may be stated as follows:

$$\sum_{k=1}^{p-1}\left(\frac{k}{p}\right)e^{2\pi ika/p} = e^{\pi i(p-1)^2/8}p^{1/2}\left(\frac{a}{p}\right), \tag{12}$$

for any integer a. For a proof of (12) the reader is referred to H. Rademacher, *Lectures in Elementary Number Theory* (New York: Blaisdell Publishing Company, 1964), pp. 88–93. Inserting (12) into (11), we obtain (7) for odd r, and the proof of Theorem 4 is complete.

Proof of Theorem 3. By Theorem 2, $h^r(\tau)$ is invariant with respect to the group $\Gamma_0(p^2)$, since r is an integer. By Theorem 10 of chapter 1, W^{-pk}, $0 \leqslant k \leqslant p-1$, form a complete set of coset representatives of $\Gamma_0(p)$ with respect to $\Gamma_0(p^2)$. Hence, by Theorem 15 of Chapter 2, $S(V\tau) = S(\tau)$, for all $V \in \Gamma_0(p)$. It is also obvious from the definition that $S(\tau)$ is regular in $\mathscr{H}$. It remains to consider the behavior of $S(\tau)$ at the parabolic cusps ∞ and 0 in the S.F.R. of $\Gamma_0(p)$ that we constructed in Chapter 1.

To get the expansion of $S(\tau)$ at ∞ we apply Theorem 4 directly. Consider first

$$h^r(\tau) = \frac{\eta^r(p^2\tau)}{\eta^r(\tau)}$$

$$= \frac{e^{\pi i r p^2 \tau/12} \prod_{n=1}^{\infty} (1 - e^{2\pi i p^2 n \tau})^r}{e^{2\pi i r \tau/12} \prod_{n=1}^{\infty} (1 - e^{2\pi i n \tau})^r}$$

$$= e^{2\pi i r v \tau}(1 + \cdots).$$

Suppose r is even. The first admissible value of n in the infinite sum on the right-hand side of (6) is $n = rv - [rv/p]p$, where $[rv/p]$ denotes the largest integer $\leqslant rv/p$. Thus we have shown that, if r is even, $S(\tau)$ has an expansion at ∞ of the required sort; that is,

$$S(\tau) = \sum_{m=-\mu}^{\infty} a_m e^{2\pi i m \tau}.$$

In fact, we have proved significantly more, namely, that if r is even and *positive*, then $S(\tau) + p^{-r/2}e^{(\pi i r/4)(1-p)}$ has a zero at ∞ of order at least $rv - [rv/p]p$, and if r is even and *negative*, then $S(\tau)$ has a pole at ∞ of order $-r$.

Suppose next that r is odd. Since $p_r(0) = 1$, the first term in the infinite sum on the right-hand side of (7) is $\left(\frac{rv}{p}\right)$. Thus if r is odd, $S(\tau)$ has the expansion at ∞:

$$(13)\qquad S(\tau) = e^{2\pi i r v \tau}(1 + \cdots) + e^{\pi i (p-1)^2/8} p^{(1-r)/2} e^{(\pi i r/4)(1-p)} \left\{\left(\frac{rv}{p}\right) + \cdots\right\},$$

which is again of the required sort.

We now turn to the expansion of $S(\tau)$ at 0. We apply the identity (10) that occurred in the proof of Theorem 4. In (10) we replace τ by $T\tau = -1/\tau$. This gives, for $1 \leqslant k \leqslant p - 1$,

$$h(W^{-pk}T\tau) = p^{-1/2}e^{(\pi i/4)(1-p)}e^{\pi i k' p/12}\left(\frac{k'}{p}\right)\frac{\eta(T\tau - k'/p)}{\eta(T\tau)}.$$

We can write $T\tau - k'/p = M_2\left(\frac{\tau + kp}{p^2}\right)$, with

$$M_2 = \begin{pmatrix} -k' & (kk'-1)/p \\ p & -k \end{pmatrix},$$

as a simple calculation shows. We have

$$\eta(T\tau) = e^{-\pi i/4}\tau^{1/2}\eta(\tau),$$

and

$$\eta\left(M_2\left(\frac{\tau + kp}{p^2}\right)\right) = v_\eta(M_2)(\tau/p)^{1/2}\eta\left(\frac{\tau + kp}{p^2}\right),$$

where, by Theorem 2 of Chapter 4,

$$\begin{aligned} v_\eta(M_2) &= \left(\frac{-k}{p}\right)^* \exp\left(\frac{\pi i}{12}\left\{-p(k' + k) + k\frac{kk' - 1}{p}(p^2 - 1) - 3p\right\}\right) \\ &= \left(\frac{-k}{p}\right)e^{(\pi i/12)(-k'p - kp - 3p)}. \end{aligned}$$

Here we have also used the fact that $p^2 - 1 \equiv 0 \pmod{24}$. Thus we obtain

$$\begin{aligned} h(W^{-pk}T\tau) = p^{-1/2}e^{(\pi i/4)(1-p)}e^{\pi i k' p/12}\left(\frac{k'}{p}\right)\left(\frac{-k}{p}\right)p^{-1/2}e^{\pi i/4} \\ \times e^{(\pi i/12)(-k'p - kp - 3p)}\frac{\eta\left(\dfrac{\tau + kp}{p^2}\right)}{\eta(\tau)}. \end{aligned}$$

Using the fact that $\left(\frac{k'}{p}\right)\left(\frac{-k}{p}\right) = \left(\frac{-1}{p}\right) = e^{(\pi i/2)(1-p)}$, we obtain, after suitable simplifications,

$$h(W^{-pk}T\tau) = p^{-1}e^{-\pi i k p/12}\eta\left(\frac{\tau + kp}{p^2}\right)\Big/\eta(\tau), \tag{14}$$

for $1 \leqslant k \leqslant p - 1$. If $k = 0$, on the other hand, we have

$$\begin{aligned} h(T\tau) &= \eta(p^2 T\tau)/\eta(T\tau) \\ &= e^{-\pi i/4}\left(\frac{\tau}{p^2}\right)^{1/2}\eta\left(\frac{\tau}{p^2}\right)\Big/e^{-\pi i/4}\tau^{1/2}\eta(\tau) \\ &= p^{-1}\eta(\tau/p^2)/\eta(\tau), \end{aligned}$$

so that (14) holds for $0 \leqslant k \leqslant p - 1$.

We apply (14) in the definition of $S(\tau)$ to find that

$$\begin{aligned}
S(T\tau) &= \sum_{k=0}^{p-1} h^r(W^{-pk}T\tau) \\
&= p^{-r}\eta^{-r}(\tau) \sum_{k=0}^{p-1} e^{-\pi i k p r/12}\eta^r\left(\frac{\tau+kp}{p^2}\right) \\
&= p^{-r}e^{-\pi i r\tau/12} \prod_{n=1}^{\infty}(1-e^{2\pi i n\tau})^{-r} \\
&\quad \times \sum_{k=0}^{p-1} e^{-\pi i k p r/12} \exp\left[\left(\frac{\pi i r}{12}\right)\left(\frac{\tau+kp}{p^2}\right)\right] \sum_{n=0}^{\infty} p_r(n)\exp\left[2\pi i n\left(\frac{\tau+kp}{p^2}\right)\right] \\
&= p^{-r}\exp\left(\frac{-2\pi i\tau}{p}\frac{vr}{p}\right)\prod_{n=1}^{\infty}(1-e^{2\pi i n\tau})^{-r} \\
&\quad \times \sum_{n=0}^{\infty} p_r(n)e^{2\pi i n\tau/p^2} \sum_{k=0}^{p-1} e^{(2\pi i/p)(n-rv)k}.
\end{aligned}$$

The finite inner sum on k

$$= \begin{cases} p & \text{if } p \mid (n-rv), \\ 0 & \text{if } p \nmid (n-rv). \end{cases}$$

Thus we obtain the following identity for all r (odd or even), which is analogous to Theorem 4:

$$(15) \qquad S(T\tau) = p^{1-r}\prod_{n=1}^{\infty}(1-e^{2\pi i n\tau})^{-r} \sum_{\substack{n=0 \\ n\equiv rv \,(\mathrm{mod}\, p)}}^{\infty} p_r(n)e^{(2\pi i\tau/p^2)(n-rv)}.$$

The first admissible value in the infinite sum on the right-hand side of (15) is $n = rv - [rv/p]p$, so that if we replace τ by $T\tau$ in (15) we find that $S(\tau)$ has the expansion at 0

$$S(\tau) = p^{1-r}p_r\left(rv - \left[\frac{rv}{p}\right]p\right)e^{-2\pi i[rv/p](T\tau)/p} + \sum_{m>-[rv/p]} a_m e^{2\pi i m(T\tau)/p},$$

an expansion of the required type. This completes the proof of Theorem 3.

We shall be applying the function $S_{p,r}(\tau)$ only in the case $r > 0$. For our applications it is important to have available the expansions of $S_{p,r}(\tau)$ at the cusps ∞ and 0. Thus we state the following theorem, whose proof is already contained in the proof of Theorem 3.

Theorem 6. Let $r > 0$. At 0 the function $S_{p,r}(\tau)$ has a pole of order at most $[rv/p]$. This pole is of order exactly $[rv/p]$ if and only if $p_r(rv - [rv/p]p) \neq 0$.

At ∞ the function $S_{p,r}(\tau)$ is regular. In addition, if r is even, then $S_{p,r}(\tau) + p^{-r/2}e^{(\pi ir/4)(1-p)}$ has a zero at ∞ of order at least $rv - [rv/p]p$.

We conclude this section with the

Proof of Lemma 5. (a) We have

$$k\left\{\frac{kk'-1}{p}\right\} + k'(k^2-1)\left\{\frac{kk'-1}{p}\right\} = -\frac{k}{p} + \frac{k'^2k^3}{p} - \frac{kk'^2}{p} + \frac{k'}{p}.$$

Thus, denoting the left-hand side of the statement of the lemma by E, we have

$$E - pk' = (k-k')\frac{p^2-1}{p} + \frac{kk'^2(k^2-1)}{p}.$$

Thus $p(E-pk') = (k-k')(p^2-1) + k'^2k(k^2-1)$. Since p is a prime > 3, $p^2 - 1 \equiv 0 \pmod{24}$. Also, since k is odd, $k(k^2-1) = (k-1)k(k+1) \equiv 0 \pmod{24}$. It follows that $p(E-pk') \equiv 0$ (mod (24). On the other hand, $(24,p) = 1$, so that $E - pk' \equiv 0 \pmod{24}$, or $E \equiv pk' \pmod{24}$, as was to be proved.

(b) Here we make free use of Lemma 1 of Chapter 4. We have

$$\left(\frac{k}{kk'-1}\right) = \left(\frac{2}{kk'-1}\right)^{\alpha}\left(\frac{k^*}{kk'-1}\right)$$

$$= (-1)^{\frac{(kk'-1)^2-1}{8}\alpha}(-1)^{\frac{k^*-1}{2}\frac{kk'-1}{2}}\left(\frac{kk'-1}{k^*}\right).$$

But

$$\left(\frac{kk'-1}{k^*}\right) = \left(\frac{-1}{k^*}\right) = (-1)^{\frac{k^*-1}{2}},$$

so that

$$\left(\frac{k}{kk'-1}\right) = (-1)^{\frac{(kk'-1)^2-1}{8}\alpha + \frac{k^*-1}{2}\frac{kk'}{2}}.$$

(c) We denote the left-hand side of the proposed congruence by E. Then we have

$$\begin{aligned} p(E - 3 + 3p - pk') &= \frac{(kk'-1)^2-1}{2}3\alpha p + 3(k^*-1)kk'p - 2p^2k \\ &\quad - 2k(kk'-1) + k'(kk'-1)(k^2-1) \\ &\quad + 3(kk'-1) + 3p^2 - p^2k'. \end{aligned} \tag{16}$$

We show first that $p(E - 3 + 3p - pk') \equiv 0 \pmod 3$. To do this it is sufficient to prove that

$$(17) \quad -2p^2k - 2k(kk' - 1) + k'(kk' - 1)(k^2 - 1) - p^2k' \equiv 0 \pmod 3.$$

If $3|k$, then the left-hand side of (17)

$$\equiv k'(1 - p^2) \equiv 0 \pmod 3.$$

If $3\nmid k$, then $k^2 - 1 \equiv 0 \pmod 3$, and the left-hand side of (17)

$$\equiv -2p^2k - 2k(kk' - 1) - p^2k' \equiv -2k - 2k' + 2k - k' \equiv 0 \pmod 3,$$

since of course $p^2 \equiv 1 \pmod 3$.

Next we show that $p(E - 3 + 3p - pk') \equiv 0 \pmod 8$. Since $p^2 \equiv 1 \pmod 8$, it is sufficient to prove that

$$\frac{(kk' - 1)^2 - 1}{2} 3\alpha p + 3(k^* - 1)kk'p - 2k - 2k(kk' - 1) + k'(kk' - 1)(k^2 - 1) + 3(kk' - 1) + 3 - k' \equiv 0 \pmod 8,$$

or, after a minor rearrangement,

$$\frac{(kk' - 1)^2 - 1}{2} 3\alpha p + 3(k^* - 1)kk'p - 2k^2k' + k'(k^3k' - k^2 - kk') + 3kk' \equiv 0 \pmod 8.$$

Since k is even, this reduces to

$$(18) \quad \frac{(kk' - 1)^2 - 1}{2} 3\alpha p + 3(k^* - 1)kk'p - k'(k^2 + kk') + 3kk' \equiv 0 \pmod 8.$$

Suppose $\alpha \geqslant 2$; then $4|k$ and (18) reduces to

$$\left(\frac{k^2k'^2}{2} - kk'\right) 3\alpha p - kk'^2 + 3kk' \equiv 0 \pmod 8.$$

If $\alpha \geqslant 3$, then actually $8|k$, and this congruence obviously holds. If $\alpha = 2$, then $4|k$ and the congruence reduces to $kk'(3 - k') \equiv 0 \pmod 8$. But this clearly holds, since either k' is even or $3 - k'$ is even.

We must show that (18) holds if $\alpha = 1$. In this case $k^* = k/2$, and $k \equiv 2 \pmod 4$. Then (18) reduces to

$$\left(\frac{k^2k'^2}{2} - kk'\right) 3p + 3\left(\frac{k}{2} - 1\right) kk'p - k'(k^2 + kk') + 3kk' \equiv 0 \pmod 8.$$

If k' is even, this in turn becomes

$$3kk'(1 - p) \equiv 0 \pmod 8,$$

which clearly holds since p is odd. Thus we may suppose that k' is odd. With k' odd, $k'^2 \equiv 1 \pmod 8$ and the desired congruence becomes

$$\left(\frac{k^2}{2} - kk'\right)3p + 3\left(\frac{k}{2} - 1\right)kk'p - k'k^2 - k + 3kk' \equiv 0 \pmod 8,$$

or

$$\frac{3pk^2}{2}(k' + 1) - 6pkk' + k(3k' - k'k - 1) \equiv 0 \pmod 8.$$

Since $2kk' - 6pkk' = 2kk'(1 - 3p) \equiv 0 \pmod 8$, this is, in turn, equivalent to

$$\frac{3pk^2}{2}(k' + 1) + k(k' - k'k - 1) \equiv 0 \pmod 8. \tag{19}$$

Since k' is odd, either $k' \equiv 1 \pmod 4$ or $k' \equiv 3 \pmod 4$. Suppose $k' \equiv 1 \pmod 4$; then since $k^2/2$ and k are even, the left-hand side of (19)

$$\equiv 3pk^2 + k(-k) \equiv k^2(3p - 1) \equiv 0 \pmod 8.$$

If $k' \equiv 3 \pmod 4$, the left-hand side of (19)

$$\equiv 6pk^2 + k(2 - 3k) \equiv k(2 - 3k) \pmod 8.$$

Since $k \equiv 2 \pmod 4$, $2 - 3k \equiv 0 \pmod 4$, so that $k(2 - 3k) \equiv 0 \pmod 8$. In either case (19) holds.

We have shown that $p(E - 3 + 3p - pk') \equiv 0 \pmod 3$ and also that $p(E - 3 + 3p - pk') \equiv 0 \pmod 8$. If follows that $p(E - 3 + 3p - pk') \equiv 0 \pmod{24}$, since $(3,8) = 1$. Since $(p,24) = 1$ it follows that $E - 3 + 3p - pk' \equiv 0 \pmod{24}$, and the proof of Lemma 5(c) is complete.

4. THE CONGRUENCE FOR $p(n)$ MODULO 11

This section is devoted to Newman's proof of 1(c) for the case $n = 1$. The congruence to be proved is

$$p(11m + 6) \equiv 0 \pmod{11}, \qquad \text{for } m = 0, 1, 2, \ldots. \tag{20}$$

Applying Theorems 3 and 6 in the case $p = 11$, $r = 10$, we find that the function $F_{11,10}(\tau) = S_{11,10}(\tau) - 11^{-5}$ is a modular function on $\Gamma_0(11)$ with a zero at ∞ of order at least 6, a pole at 0 of order at most 4, and regular in $\mathscr{H}$. On the other hand, Theorem 1 with $p = 11$, $r = 12$, implies that

$\Phi_{11,12}(\tau)$ is a modular function (since r is even) on $\Gamma_0(11)$, regular and nonzero in $\mathscr{H}$, with a zero at ∞ of order 5 and a pole at 0 of order 5. Therefore, the function $f(\tau) = F_{11,10}(\tau)/\Phi_{11,12}(\tau)$ is a modular function on $\Gamma_0(11)$, regular in $\mathscr{H}$, with a zero of order at least one at ∞ and a zero of order at least one at 0. By Theorem 7 of Chapter 2, $f(\tau)$ is constant, and, in this case, $f(\tau) \equiv 0$. Thus we conclude that $F_{11,10}(\tau) \equiv 0$. By (6) of Theorem 4, therefore, if $x = e^{2\pi i\tau}$,

$$h_{11}^{10}(\tau) - 11^{-4} \prod_{n=1}^{\infty} (1 - x^n)^{-10} \sum_{\substack{n=0 \\ n \equiv 6 \pmod{11}}}^{\infty} p_{10}(n)x^n \equiv 0. \tag{21}$$

But

$$h_{11}^{10}(\tau) = \{\eta(11^2\tau)/\eta(\tau)\}^{10}$$

$$= x^{50} \frac{\prod_{n=1}^{\infty} (1 - x^{11^2 n})^{10}}{\prod_{n=1}^{\infty} (1 - x^n)^{10}},$$

so that (21) becomes, after a rearrangement,

$$11^4 x^{50} \sum_{n=0}^{\infty} p_{10}(n)x^{11^2 n} = \sum_{\substack{n=0 \\ n \equiv 6 \pmod{11}}}^{\infty} p_{10}(n)x^n. \tag{22}$$

Here we have used the identity

$$\prod_{n=1}^{\infty} (1 - x^{11^2 n})^{10} = \sum_{n=0}^{\infty} p_{10}(n)x^{11^2 n}.$$

A comparison of coefficients on both sides of (22) yields the following result.

Theorem 7. If t is an integer $\geqslant 0$, we have $p_{10}(11t + 6) = 11^4 p_{10}((t - 4)/11)$, where we define $p_{10}(\alpha) = 0$ unless α is a nonnegative integer.

From Theorem 7 we readily derive

Theorem 8. If $R \equiv 10 \pmod{11}$ and $n \equiv 6 \pmod{11}$, then $p_R(n) \equiv 0 \pmod{11}$.

Proof. Put $R = 11Q + 10$. Then

$$\sum_{n=0}^{\infty} p_R(n)x^n = \prod_{n=1}^{\infty} (1 - x^n)^R = \prod_{n=1}^{\infty} (1 - x^n)^{11Q+10}$$

$$= \prod_{n=1}^{\infty} (1 - x^n)^{11Q}(1 - x^n)^{10}$$

$$\equiv \prod_{n=1}^{\infty} (1 - x^{11n})^Q(1 - x^n)^{10} \pmod{11},$$

where the congruence is meant coefficient-wise.

Thus

$$\sum_{n=0}^{\infty} p_R(n)x^n \equiv \sum_{n=0}^{\infty} p_Q(n)x^{11n} \sum_{n=0}^{\infty} p_{10}(n)x^n$$

$$\equiv \sum_{n=0}^{\infty} p_Q\left(\frac{n}{11}\right)x^n \sum_{n=0}^{\infty} p_{10}(n)x^n \pmod{11}.$$

If we now compare cofficients on both sides of this congruence we obtain

$$p_R(n) \equiv \sum_{j=0}^{n} p_Q\left(\frac{j}{11}\right)p_{10}(n-j) \pmod{11},$$

or

$$p_R(n) \equiv \sum_{0 \leqslant j \leqslant n/11} p_Q(j)p_{10}(n-11j) \pmod{11}.$$

By Theorem 7, $p_{10}(n-11j) \equiv 0 \pmod{11}$, if $n \equiv 6 \pmod{11}$. It follows that for $n \equiv 6 \pmod{11}$, $p_R(n) \equiv 0 \pmod{11}$, and the proof is complete.

To prove (20) we simply take the case $R = -1$ in Theorem 8. Since $p_{-1}(n) = p(n)$, (20) follows.

5. NEWTON'S FORMULA

The formula in question is one relating the coefficients of a polynomial to the power sums of the roots. This formula will be important to us at several key junctures in the remainder of this chapter and in Chapter 8. Suppose we consider the polynomial

$$f(x) = x^q - p_1x^{q-1} + p_2x^{q-2} - \cdots + (-1)^q p_q,$$

with roots $\varphi_1,\ldots,\varphi_q$. Newton's formula can then be stated in the following way.

Theorem 9. For h a positive integer put $S_h = \Sigma_{i=1}^{q} \varphi_i^h$. Then if $1 \leqslant h \leqslant q$, we have

$$S_R - p_1S_{h-1} + p_2S_{h-2} - \cdots + (-1)^{h-1}p_{h-1}S_1 + (-1)^h p_h h = 0, \tag{23}$$

and if $h > q$, we have

$$S_h - p_1S_{h-1} + p_2S_{h-2} - \cdots + (-1)^q p_q S_{h-q} = 0. \tag{24}$$

Proof. Suppose $h \leqslant q$ and let $1 \leqslant k \leqslant h$. Then, as is well known,

$$p_k = \sum_{1 \leqslant i_1 < i_2 < \cdots < i_k \leqslant h} \varphi_{i_1}\varphi_{i_2}\cdots\varphi_{i_k}.$$

Thus, for $1 \leqslant k \leqslant h-1$,

$$\begin{aligned} p_k S_{h-k} &= \left(\sum_{1 \leqslant i_1 < \cdots < i_k \leqslant h} \varphi_{i_1}\varphi_{i_2}\cdots\varphi_{i_k} \right)\left(\sum_{i=1}^{q} \varphi_i^{h-k} \right) \\ &= \sum\nolimits^* \varphi_{i_1}\cdots\varphi_{i_{k-1}}\varphi_{i_k}^{h-k+1} \\ &\quad + \sum\nolimits^* \varphi_{i_1}\cdots\varphi_{i_k}\varphi_{i_{k+1}}^{h-k}, \end{aligned}$$

where Σ^* indicates that we are to sum over all possible values of the subscripts, subject to the condition that no two subscripts are to have the same value; that is, $i_a \neq i_b$ if $a \neq b$. In particular, for $k = 1$ we have

$$\begin{aligned} p_1 S_{h-1} &= \sum\nolimits^* \varphi_{i_1}^h + \sum\nolimits^* \varphi_{i_1}\varphi_{i_2}^{h-1} \\ &= S_h + \sum\nolimits^* \varphi_{i_1}\varphi_{i_2}^{h-1}, \end{aligned}$$

and for $k = h-1$,

$$p_{h-1}S_1 = \sum\nolimits^* \varphi_{i_1}\cdots\varphi_{i_{h-2}}\varphi_{i_{h-1}}^2 + hp_h.$$

If we form the alternating sum on k from $k = 1$ to $k = h-1$, (23) follows.

Suppose $h > q$. Since φ_i is a root of $f(x)$, we have

$$\varphi_i^q - p_1\varphi_i^{q-1} + \cdots + (-1)^q p_q = 0, \qquad \text{for } 1 \leqslant i \leqslant q.$$

Multiply both sides of this by φ_i^{h-q} and sum on i. This gives (24), and the proof is complete.

6. THE MODULAR EQUATION FOR THE PRIME 5

The equation we are to prove is

$$\Phi(\tau) = h(\tau)\{5^2h(\tau)^4 + 5^2h(\tau)^3 + 5\cdot 3h(\tau)^2 + 5h(\tau) + 1\}, \tag{25}$$

where $\Phi(\tau) = \Phi_{5,6}(\tau) = \{\eta(5\tau)/\eta(\tau)\}^6$ and $h(\tau) = h_5(\tau) = \eta(25\tau)/\eta(\tau)$. We also write

$$S_r(\tau) = S_{5,r}(\tau) = \sum_{k=0}^{4} h_5^r(W^{-5k}\tau) = \sum_{k=0}^{4} h^r(W^{-5k}\tau).$$

Consider the polynomial

$$\prod_{k=0}^{4} \{(u - h(W^{-5k}\tau)\} = u^5 - c_1u^4 + c_2u^3 - c_3u^2 + c_4u - c_5,$$

where the c_k are functions of τ. Clearly $h(\tau)$ is a root of this polynomial, so that we have

$$h(\tau)^5 - c_1h(\tau)^4 + c_2h(\tau)^3 - c_3h(\tau)^2 + c_4h(\tau) - c_5 = 0. \tag{26}$$

We now need only determine the c_k as functions of τ. As a first step we determine the functions $S_r(\tau)$, $1 \leqslant r \leqslant 5$. By Theorem 6, $S_r(\tau)$ is regular at ∞ and at 0 has a pole of order $\leqslant [r/5]$. For $1 \leqslant r \leqslant 4$, $[r/5] = 0$, so that in these cases $S_r(\tau)$ has no pole. Thus by Theorem 7 of Chapter 2, $S_r(\tau)$, for $1 \leqslant r \leqslant 4$, is constant. By Theorem 6 again, $S_r(\tau) = -5^{-r/2}$, for $r = 2$ and 4, whereas by equation (13), $S_r(\tau) = -5^{(1-r)/2}\left(\frac{r}{5}\right)$, for $r = 1$ and 3. The function $S_5(\tau)$ has a pole at 0 of order $\leqslant [5/5] = 1$. Since $p_5(5-5) = p_5(0) = 1 \neq 0$, it follows from Theorem 6 that the pole at 0 is of order exactly 1. Since the Legendre symbol $\left(\frac{5}{5}\right) = 0$, it follows from equation (13) that $S_5(\tau)$ has a zero at ∞ of order at least 1. Thus by Theorem 1 the function $f(\tau) = S_5(\tau)/\Phi(\tau)$ is a modular function on $\Gamma_0(5)$, regular in $\mathscr{H}$, and also regular at the parabolic cusps ∞ and 0. By Theorem 7 of Chapter 2, $f(\tau)$ is a constant. Since the expansion of $\Phi(\tau)$ at 0 begins with the term $5^{-3}e^{-(2\pi i/5)(-1/\tau)}$ and that of $S_5(\tau)$ at 0 with the term $5^{-4}e^{-(2\pi i/5)(-1/\tau)}$, it follows that $f(\tau) = 5^{-1}$, so that $S_5(\tau) = 5^{-1}\Phi(\tau)$. In summary, we have $S_1(\tau) = -1$, $S_2(\tau) = -5^{-1}$, $S_3(\tau) = 5^{-1}$, $S_4(\tau) = -5^{-2}$, $S_5(\tau) = 5^{-1}\Phi(\tau)$.

From the original product definition it is clear that the roots of the polynomial under discussion are precisely $h(W^{-5k}\tau)$, $0 \leqslant k \leqslant 4$. Hence by Newton's formula (Theorem 9), we can now determine the c_k that occur in (26). We have

$$c_1 = S_1 = -1$$

$$c_2 = \frac{c_1 S_1 - S_2}{2} = \frac{3}{5}$$

$$c_3 = \frac{c_2 S_1 - c_1 S_2 + S_3}{3} = -\frac{1}{5}$$

$$c_4 = \frac{c_3 S_1 - c_2 S_2 + c_1 S_3 - S_4}{4} = \frac{1}{5^2}$$

$$c_5 = \frac{c_4 S_1 - c_3 S_2 + c_2 S_3 - c_1 S_4 + S_5}{5} = \frac{\Phi(\tau)}{5^2}.$$

Inserting these values into (26), we get

$$h(\tau)^5 + h(\tau)^4 + \frac{3}{5}h(\tau)^3 + \frac{1}{5}h(\tau)^2 + \frac{1}{5^2}h(\tau) - \frac{1}{5^2}\Phi(\tau) = 0,$$

and, with a minor rearrangement, (25) follows.

7. THE MODULAR EQUATION FOR THE PRIME 7

The equation in question in this instance is

$$\begin{aligned}&\Phi(\tau)^2 - \Phi(\tau)\{7^2h(\tau)^3 + 7\cdot 5h(\tau)^2 + 7h(\tau)\}\\ &-\{7^3h(\tau)^7\\ &+ 7^3h(\tau)^6 + 3\cdot 7^2h(\tau)^5 + 7^2h(\tau)^4 + 3\cdot 7h(\tau)^3 + 7h(\tau)^2 + h(\tau)\} = 0,\end{aligned} \tag{27}$$

where $\Phi(\tau) = \Phi_{7,4}(\tau) = \{\eta(7\tau)/\eta(\tau)\}^4$ and $h(\tau) = h_7(\tau) = \eta(49\tau)/\eta(\tau)$. Also put

$$S_r(\tau) = S_{7,r}(\tau) = \sum_{k=0}^{6} h^r(W^{-7k}\tau).$$

Proceeding as in Section 6, we consider the polynomial

$$\prod_{k=0}^{6}\{u - h(W^{-7k}\tau)\} = u^7 - c_1u^6 + c_2u^5 - c_3u^4 + c_4u^3 - c_5u^2 + c_6u - c_7,$$

where the c_k are functions of τ. Since $h(\tau)$ is a root of this polynomial, we have

$$\begin{aligned}&h(\tau)^7 - c_1h(\tau)^6 + c_2h(\tau)^5 - c_3h(\tau)^4 + c_4h(\tau)^3 - c_5h(\tau)^2\\ &\quad + c_6h(\tau) - c_7 = 0.\end{aligned} \tag{28}$$

By Theorem 6, $S_r(\tau)$ is regular at ∞ and has a pole at 0 of order $\leqslant[2r/7]$. For $1 \leqslant r \leqslant 3$, $[2r/7] = 0$, so that $S_r(\tau)$ has no pole and, by Theorem 7 of Chapter 2, is constant. By equation (13), $S_1(\tau) = -1$ and $S_3(\tau) = -7^{-1}$. By Theorem 6, $S_2(\tau) = 7^{-1}$. Furthermore, $[12/7] = 1$, but $p_6(5) = 0$, so that $S_6(\tau)$ has a pole at 0 of order <1. That is, $S_6(\tau)$ also is constant. By Theorem 6, $S_6(\tau) = 7^{-3}$.

By equation (6) of Theorem 4 we know that the first term in the expansion of $S_4(\tau) + 7^{-2}$ at ∞ is $7^{-1}p_4(1)e^{2\pi i\tau} = -4\cdot 7^{-1}e^{2\pi i\tau}$. By Theorem 1 the expansion of $\Phi(\tau)$ at ∞ begins with the term $e^{2\pi i\tau}$. Furthermore, $S_4(\tau) + 7^{-2}$ has a pole of order $\leqslant 1$ at 0 and $\Phi(\tau)$ has a pole of order exactly 1 at 0. Thus $(S_4(\tau) + 7^{-2})/\Phi(\tau)$, once more by Theorem 7 of Chapter 2, is a constant. In fact, $S_4(\tau) + 7^{-2} = -4\cdot 7^{-1}\Phi(\tau)$. By equation (7) of Theorem 4 the first term in the expansion of $S_5(\tau) - 7^{-2}$ at ∞ is

$$\begin{aligned}-7^{-2}\left(\frac{9}{7}\right)p_5(1)e^{2\pi i\tau} - 7^{-2}\left(\frac{10}{7}\right)p_{-5}(1)e^{2\pi i\tau} &= (5\cdot 7^{-2} + 5\cdot 7^{-2})e^{2\pi i\tau}\\ &= 10\cdot 7^{-2}e^{2\pi i\tau}.\end{aligned}$$

Since $S_5 - 7^{-2}$ has a pole of order $\leqslant 1$ at 0, the same argument that was applied above to $S_4(\tau)$ yields $S_5(\tau) - 7^{-2} = 10\cdot 7^{-2}\Phi(\tau)$. Finally, we consider $S_7(\tau)$. Since $p_7(0) = 1 \neq 0$, by Theorem 6, $S_7(\tau)$ has a pole at 0 of order exactly 2. By equation (7) of Theorem 4 the expansion of $S_7(\tau)$ at the cusp ∞ begins with

$$7^{-3}\left\{\left(\frac{13}{7}\right)p_7(1)e^{2\pi i\tau} + \left(\frac{12}{7}\right)p_7(2)e^{4\pi i\tau}\right\}\{1 + p_{-7}(1)e^{2\pi i\tau}\}.$$

A simple calculation shows that $p_7(1) = -7$, $p_7(2) = 14$, and $p_{-7}(1) = 7$. Since $\left(\frac{13}{7}\right) = \left(\frac{12}{7}\right) = -1$, the expansion of $S_7(\tau)$ at ∞ begins with

$$7^{-3}(7e^{2\pi i\tau} - 14e^{4\pi i\tau})(1 + 7e^{2\pi i\tau}),$$

so that the first two terms are $7^{-2}e^{2\pi i\tau} + 5\cdot 7^{-2}e^{4\pi i\tau}$. In the course of the proof of Theorem 1 we derived the following expansion for $\Phi(\tau) = \Phi_{7,4}(\tau)$:

$$\Phi(\tau) = e^{2\pi i\tau}\prod_{m=1}^{\infty}(1 - e^{2\pi i m\cdot 7\tau})^4\prod_{m=1}^{\infty}(1 - e^{2\pi i m\tau})^{-4}.$$

From this we readily see that the first two terms in the expansion of $\Phi(\tau)$ at ∞ are $e^{2\pi i\tau} + 4e^{4\pi i\tau}$. Thus the function $S_7 - 7^{-2}\Phi(\tau)$ has a pole at 0 of order exactly 2 and has an expansion at ∞ beginning with the term $7^{-2}e^{4\pi i\tau}$. Applying Theorem 7 of Chapter 2 in the by-now-familiar way, we find that $\{S_7 - 7^{-2}\Phi(\tau)\}/\Phi^2(\tau) = 7^{-2}$, or $S_7 = 7^{-2}\Phi^2(\tau) + 7^{-2}\Phi(\tau)$.

In summary, we have $S_1(\tau) = -1$, $S_2(\tau) = 7^{-1}$, $S_3(\tau) = -7^{-1}$, $S_4(\tau) = -4\cdot 7^{-1}\Phi(\tau) - 7^{-2}$, $S_5(\tau) = 10\cdot 7^{-2}\Phi(\tau) + 7^{-2}$, $S_6(\tau) = 7^{-3}$, $S_7(\tau) = 7^{-2}\Phi^2(\tau) + 7^{-2}\Phi(\tau)$.

We now apply Newton's formula to determine the c_k that occur in (28). Since it is clear that the roots of the polynomial $\Pi_{k=0}^{6}\{u - h(W^{-7k}\tau)\}$ are precisely $h(W^{-7k}\tau)$, $0 \leqslant k \leqslant 6$, we have

$$c_1 = S_1 = -1$$

$$c_2 = \frac{c_1S_1 - S_2}{2} = \frac{3}{7}$$

$$c_3 = \frac{c_2S_1 - c_1S_2 + S_3}{3} = -\frac{1}{7}$$

$$c_4 = \frac{c_3S_1 - c_2S_2 + c_1S_3 - S_4}{4} = \frac{1}{7}\Phi + \frac{3}{7^2}$$

$$c_5 = \frac{c_4S_1 - c_3S_2 + c_2S_3 - c_1S_4 + S_5}{5} = -\frac{5}{7^2}\Phi - \frac{1}{7^2}$$

$$c_6 = \frac{c_5S_1 - c_4S_2 + c_3S_3 - c_2S_4 + c_1S_5 - S_6}{6} = \frac{1}{7^2}\Phi + \frac{1}{7^3}$$

$$c_7 = \frac{c_6S_1 - c_5S_2 + c_4S_3 - c_3S_4 + c_2S_5 - c_1S_6 + S_7}{7} = \frac{1}{7^3}\Phi^2.$$

Putting these values into (28), we obtain

$$h(\tau)^7 + h(\tau)^6 + \frac{3}{7}h(\tau)^5 + \frac{1}{7}h(\tau)^4 + \left(\frac{1}{7}\Phi + \frac{3}{7^2}\right)h(\tau)^3$$
$$+ \left(\frac{5}{7^2}\Phi + \frac{1}{7^2}\right)h(\tau)^2 + \left(\frac{1}{7^2}\Phi + \frac{1}{7^3}\right)h(\tau) - \frac{1}{7^3}\Phi^2 = 0.$$

Multiplying through by 7^3 and collecting on powers of Φ, we arrive at (27).

Chapter 8

PROOF OF THE RAMANUJAN CONGRUENCES FOR POWERS OF 5 AND 7

1. PRELIMINARIES

In this chapter we give Atkin's revised version of Watson's proof of (1a) and (1b) of Chapter 7. In fact, as we shall see presently, Atkin's argument proves somewhat more than (1a) and (1b). In the first half of the chapter we confine ourselves to (1a), that is, to the prime 5, and reserve the proof of (1b) for the second half.

It will be convenient to introduce some new notation. For n a positive integer, define

(1) $$l_{2n-1} = (19 \cdot 5^{2n-1} + 1)/24, \qquad l_{2n} = (23 \cdot 5^{2n} + 1)/24.$$

It is not hard to verify that l_{2n-1} and l_{2n} are integers, and, in fact, l_n is the smallest positive integral solution of $24x \equiv 1 \pmod{5^n}$. Hence if x is any positive integral solution, we have $24(x - l_n) \equiv 0 \pmod{5^n}$, or $x \equiv l_n \pmod{5^n}$. Thus every positive integral solution of $24x \equiv 1 \pmod{5^n}$ has the form $x = m5^n + l_n$, with m a nonnegative integer. Thus (1a) of Chapter 7 can be rewritten as

(2) $$p(5^n m + l_n) \equiv 0 \pmod{5^n}, \qquad m \geqslant 0,$$

for every positive integer n. In fact, we shall also prove that, for every positive integer n,

(3) $$p(l_n) \not\equiv 0 \pmod{5^{n+1}},$$

so that the power of 5 occurring in (2) is the best possible.

We introduce several functions defined in the unit disc, $|x| < 1$. These are

$$F(x) = \prod_{m=1}^{\infty} (1 - x^m)$$

$$\phi(x) = xF(x^{25})/F(x)$$

$$L_{2n-1}(x) = F(x^5) \sum_{m=0}^{\infty} p(5^{2n-1}m + l_{2n-1})x^{m+1}$$

$$L_{2n}(x) = F(x) \sum_{m=0}^{\infty} p(5^{2n}m + l_{2n})x^{m+1},$$

where n is a positive integer. It is clear that for $\tau \in \mathscr{H}$, $\eta(\tau) = e^{\pi i\tau/12}F(e^{2\pi i\tau})$ and $\phi(e^{2\pi i\tau}) = \eta(25\tau)/\eta(\tau) = h_5(\tau) = h(\tau)$. Also the absolute convergence in the unit disc of the series defining $L_n(x)$ follows from Proposition 1 of Chapter 6.

We introduce one further definition for the purpose of rewriting (2) and (3) in a more convenient form. If a is a nonzero integer, let $\pi(a)$ be the highest power of 5 dividing a, so that

$$5^{\pi(a)}|a, \qquad 5^{\pi(a)+1} \nmid a.$$

Also put $\pi(0) = \infty$ and $\pi(a/5^c) = -c$, if a and c are integers such that $(a,5) = 1$ and $c > 0$. With these definitions we have

$$(4) \qquad \begin{cases} \pi(ab) = \pi(a) + \pi(b), & \\ \pi(a + b) = \min\{\pi(a),\pi(b)\} & \text{if } \pi(a) \neq \pi(b), \\ \pi(a + b) \geqslant \min\{\pi(a),\pi(b)\} & \text{if } \pi(a) = \pi(b), \end{cases}$$

where a and b are any rational numbers. Finally, if $f(x) = \Sigma_{n=N}^{\infty} \lambda_n x^n$, with N an integer and λ_n an integer for all $n \geqslant N$, define

$$\pi\{f(x)\} = \min_{n \geqslant N} \{\pi(\lambda_n)\}.$$

Remark. If $f_1(x) = \Sigma_{n=N}^{\infty} \lambda_n x^n$ and $f_2(x) = \Sigma_{n=M}^{\infty} \mu_n x^n$, with λ_n and μ_n integers and $\pi(\mu_M) = 0$, then $\pi\{f_1(x)f_2(x)\} = \pi\{f_1(x)\}$. To prove this observe that $f_1(x)f_2(x) = \Sigma_{n=N+M}^{\infty} \nu_n x^n$, where $\nu_n = \Sigma_{j+k=n} \lambda_j \mu_k$. Hence by (4) we have

$$\pi(\lambda_n) \geqslant \min_{j+k=n} \{\pi(\lambda_j \mu_k)\} \geqslant \min_j \{\pi(\lambda_j)\} = \pi\{f_1(x)\}.$$

On the other hand, suppose $\pi\{f_1(x)\} = \pi(\lambda_t)$, where $\pi(\lambda_n) > \pi(\lambda_t)$ for $N \leqslant n \leqslant t - 1$. Then $\nu_{t+M} = \lambda_t \mu_M + \nu'_{t+M}$, where $\nu'_{t+M} = \Sigma_{j=N}^{t-1} \lambda_j \mu_{M+t-j}$. By (4) $\pi(\nu'_{t+M}) > \pi(\lambda_t)$, since $\pi(\lambda_n) > \pi(\lambda_t)$ for $N \leqslant n \leqslant t - 1$. Thus, again by (4), $\pi(\nu_{t+M}) = \pi(\lambda_t) = \pi(f_1(x)\}$. Thus, in fact, $\pi\{f_1(x)f_2(x)\} = \pi\{f_1(x)\}$.

Proposition 1. The following assertion is equivalent to (2) and (3) together, for each positive integer n:

$$\pi\{L_n(x)\} = \pi\{p(l_n)\} = n. \tag{5}$$

Proof. From the definitions we have

$$L_{2n-1}(x) = F(x^5) \sum_{m=0}^{\infty} p(5^{2n-1}m + l_{2n-1})x^{m+1}$$

$$L_{2n}(x) = F(x) \sum_{m=0}^{\infty} p(5^{2n}m + l_{2n})x^{m+1}.$$

Since the power series $F(x)$ and $F(x^5)$ both begin with the term 1, it follows from the preceding remark that

$$\pi\{L_n(x)\} = \pi\left\{\sum_{m=0}^{\infty} p(5^n m + l_n)x^{m+1}\right\},$$

for all n. Thus (5) is equivalent to

$$n = \pi\{p(l_n)\} = \pi\left\{\sum_{m=0}^{\infty} p(5^n m + l_n)x^{m+1}\right\}, \qquad \text{for all } n. \tag{6}$$

It is obvious that (6) is equivalent to (2) and (3) taken together, and the proof is complete.

We next define the linear operator $U_5 = U$ in the following way. If $f(x) = \Sigma_{n=N}^{\infty} \lambda_n x^n$, with N any integer, then†

$$U\{f(x)\} = \sum_{5n \geqslant N} \lambda_{5n} x^n. \tag{7}$$

Lemma 2. Let $f_1(x) = \Sigma_{n=N}^{\infty} \lambda_n x^n$ and $f_2(x) = \Sigma_{n=M}^{\infty} \mu_n x^n$. Then the following hold.

(a) $$U\{k_1 f_1(x) + k_2 f_2(x)\} = k_1 U\{f_1(x)\} + k_2 U\{f_2(x)\},$$

for any real numbers k_1 and k_2.

(b) $$U\{f_2(x^5) f_1(x)\} = f_2(x) U\{f_1(x)\}.$$

(c) If w is a primitive fifth root of unity, then $5U\{f_1(x)\} = \Sigma_{r=0}^{4} f_1(w^r x^{1/5})$.

Proof. The assertion (a), which says that U is linear, is trivial.

(b) $$U\{f_2(x^5) f_1(x)\} = U\left\{\sum_{m \geqslant M} \mu_m x^{5m} \sum_{n \geqslant N} \lambda_n x^n\right\}$$

$$= U\left\{\sum_{j=5M+N}^{\infty} \nu_j x^j\right\},$$

† U_5 is actually a classical Hecke operator.

where $\nu_j = \Sigma_{5m+n=j}\,\mu_m\lambda_n$. We have $\nu_{5j} = \Sigma_{5m+n=5j}\,\mu_m\lambda_n = \Sigma_{5m+5k=5j}\,\mu_m\lambda_{5k}$, so that

$$U\{f_2(x^5)f_1(x)\} = \sum_{5j \geqslant 5M+N} \nu_{5j}x^j$$

$$= \sum_{5j \geqslant 5M+N} x^j\left(\sum_{5m+5k=5j} \mu_m\lambda_{5k}\right)$$

$$= \sum_{5j \geqslant 5M+N} x^j\left(\sum_{m+k=j} \mu_m\lambda_{5k}\right)$$

$$= \left(\sum_{m \geqslant M} \mu_m x^m\right)\left(\sum_{5k \geqslant N} \lambda_{5k}x^k\right) = f_2(x)\,U\{f_1(x)\}.$$

(c)
$$\sum_{r=0}^{4} f_1(w^r x^{1/5}) = \sum_{r=0}^{4}\sum_{n \geqslant N} \lambda_n (w^r x^{1/5})^n$$

$$= \sum_{n \geqslant N} \lambda_n x^{n/5}\left(\sum_{r=0}^{4} w^{rn}\right).$$

Since w is a primitive fifth root of unity, we have $\Sigma_{r=0}^{4}\, w^{rn} = 0$, if n is not a multiple of 5, and $\Sigma_{r=0}^{4}\, w^{rn} = 5$, if n is a multiple of 5. Thus the sum becomes

$$5\sum_{5k \geqslant N} \lambda_{5k}x^k = 5U\{f_1(x)\}.$$

Lemma 3. If n is a positive integer we have

(a)
$$U\{L_{2n-1}(x)\} = L_{2n}(x)$$

and

(b)
$$U\{\phi(x)L_{2n}(x)\} = L_{2n+1}(x).$$

Proof. (a) By Lemma 2(b),

$$U\{L_{2n-1}(x)\} = F(x)U\left\{\sum_{m=0}^{\infty} p(5^{2n-1}m + l_{2n-1})x^{m+1}\right\}$$

$$= F(x)\sum_{m=0}^{\infty} p\{5^{2n-1}(5m+4) + l_{2n-1}\}x^{m+1}$$

$$= F(x)\sum_{m=0}^{\infty} p(5^n m + l_{2n})x^{m+1} = L_{2n}(x),$$

since it is clear from (1) that $4 \cdot 5^{2n-1} + l_{2n-1} = l_{2n}$.

(b) Again by Lemma 2(b),

$$\begin{aligned} U\{\phi(x)L_{2n}(x)\} &= U\left\{F(x^{25})\sum_{m=0}^{\infty} p(5^{2n}m + l_{2n})x^{m+2}\right\} \\ &= F(x^5)U\left\{\sum_{m=0}^{\infty} p(5^{2n}m + l_{2n})x^{m+2}\right\} \\ &= F(x^5)\sum_{m=0}^{\infty} p\{5^{2n}(5m+3) + l_{2n}\}x^{m+1} \\ &= F(x^5)\sum_{m=0}^{\infty} p(5^{2n+1}m + l_{2n+1})x^{m+1} \\ &= L_{2n+1}(x), \end{aligned}$$

since $3 \cdot 5^{2n} + l_{2n} = l_{2n+1}$.

2. APPLICATION OF THE MODULAR EQUATION

In Chapter 7, Section 6, the modular equation for the prime 5 was proved in the form

$$\Phi(\tau) = h(\tau)\{5^2h(\tau)^4 + 5^2h(\tau)^3 + 5 \cdot 3h(\tau)^2 + 5h(\tau) + 1\},$$

where $\Phi(\tau) = \{\eta(5\tau)/\eta(\tau)\}^6$ and $h(\tau) = \eta(25\tau)/\eta(\tau)$. After an elementary rearrangement, this becomes

(8) $$h(\tau)^5 = \Phi(5\tau)\{5^2h(\tau)^4 + 5^2h(\tau)^3 + 5 \cdot 3h(\tau)^2 + 5h(\tau) + 1\}.$$

We have already observed that with $x = e^{2\pi i\tau}$, $h(\tau) = \phi(x)$, where $\phi(x) = xF(x^{25})/F(x)$. In the same way, if $g(x) = xF^6(x^5)/F^6(x)$, then $g(x) = \Phi(\tau)$ and $g(x^5) = \Phi(5\tau)$. Thus we obtain the modular equation in the form in which we shall use it,

(9) $$\phi^5(y) = g(x^5)\{5^2\phi^4(x) + 5^2\phi^3(x) + 3 \cdot 5\phi^2(x) + 5\phi(x) + 1\}.$$

Consider the polynomial

(10) $$u^5 - g(x^5)\{5^2u^4 + 5^2u^3 + 3 \cdot 5u^2 + 5u + 1\}.$$

By (9), $\phi(x)$ is a root of (10). Furthermore, if w is a primitive fifth root of unity, $\phi(w^kx)$, $k = 0, 1, 2, 3, 4$, are all roots of (10), since $g(x^5)$ is unchanged if we replace x by w^kx, and since (9) holds identically in x, for $|x| < 1$. On the other hand,

$$\begin{aligned} \phi(w^kx) &= w^kxF(x^{25})/F(w^kx) \\ &= w^kx + \text{terms of higher order in } x. \end{aligned}$$

Since w^k, $k = 0, 1, 2, 3, 4$, are all distinct, it follows that $\phi(w^k x)$, $0 \leqslant k \leqslant 4$, are all distinct, and are thus the five distinct roots of (10). From this we conclude that the five distinct roots of the polynomial

$$u^5 - g(x)\{5^2u^4 + 5^2u^3 + 3 \cdot 5u^2 + 5u + 1\} \tag{11}$$

are $u = \phi(w^k x^{1/5})$, $k = 0, 1, 2, 3, 4$.

Using the notation of Newton's formula, we let S_r be the sum of the rth power of the roots of (11); that is,

$$S_r = \sum_{k=0}^{4} \phi^r(w^k x^{1/5}) = 5U\{\phi^r(x)\},$$

by Lemma 2(c). We now apply Newton's formula to prove

Lemma 4. Let r be a positive integer. Then S_r is a polynomial in g of the form $S_r = \Sigma_{p=1}^{\infty} a_{rp} g^p$, where a_{rp} is an integer divisible by 5, such that $\pi(a_{rp}) \geqslant \left[\dfrac{5p - r + 1}{2}\right]$ and $a_{rp} = 0$ unless $\left[\dfrac{r+4}{5}\right] \leqslant p \leqslant r$.

Proof. By Newton's formula we have

$$\begin{aligned} S_1 &= 5^2 g \\ S_2 &= 5^4 g^2 + 2 \cdot 5^2 g \\ S_3 &= 5^6 g^3 + 3 \cdot 5^4 g^2 + 9 \cdot 5g \\ S_4 &= 5^8 g^4 + 4 \cdot 5^6 g^3 + 22 \cdot 5^3 g^2 + 4 \cdot 5g \\ S_5 &= 5^{10} g^5 + 5 \cdot 5^8 g^4 + 40 \cdot 5^5 g^3 + 20 \cdot 5^3 g^2 + 5g, \end{aligned} \tag{12}$$

and also, for $r > 5$,

$$S_r = 5^2 g S_{r-1} + 5^2 g S_{r-2} + 3 \cdot 5 g S_{r-3} + 5 g S_{r-4} + g S_{r-5}. \tag{13}$$

All the statements in the lemma follow trivially from (12) and (13), with the exception of the inequality $\pi(a_{rp}) \geqslant \left[\dfrac{5p - r + 1}{2}\right]$. The proof of this inequality proceeds by strong induction on r. From (12) the inequality holds when $1 \leqslant r \leqslant 5$. Also, if $r > 5$ and $p = 1$, then $a_{rp} = 0$, so that the inequality is trivially valid. Suppose then the inequality holds for all $r < R$, with $R > 5$. By (13) it follows that, for $p > 1$,

$$a_{R,p} = 5^2 a_{R-1,p-1} + 5^2 a_{R-2,p-1} + 3 \cdot 5 a_{R-3,p-1} + 5 a_{R-4,p-1} + a_{R+5,p-1}.$$

Hence it follows from (4) that for $p > 1$,

$$\pi(a_{R,p}) \geqslant \min\left\{\left[\frac{5(p-1)-R+6}{2}\right], \left[\frac{5(p-1)-R+7}{2}\right]\right\}$$
$$= \left[\frac{5(p-1)-R+6}{2}\right] = \left[\frac{5p-R+1}{2}\right].$$

Thus the inequality holds for $r = R$ and, by induction, for all positive integers r.

Lemma 5. Suppose k is a nonnegative integer. We have the following:

(a) $U(g^k(x)) = S_{6k}/5g^k(x)$.

(b) $U\{\phi(x)g^k(x)\} = S_{6k+1}/5g^k(x)$.

(c) Both $U\{g^k(x)\}$ and $U\{\phi(x)g^k(x)\}$ are polynomials in $g(x)$ with integral coefficients. If $k \geqslant 1$ the polynomial has no constant term.

Proof.

(a) $$U\{g^k(x)\} = U\{x^k F^{6k}(x^5)/F^{6k}(x)\}$$
$$= U\left\{\frac{F^{6k}(x^5)}{F^{6k}(x^{25})}\frac{\phi^{6k}(x)}{x^{5k}}\right\} = \frac{F^{6k}(x)}{F^{6k}(x^5)}\frac{1}{x^k}U\{\phi^{6k}(x)\},$$

by Lemma 2(b). But we have already observed that $U\{\phi^{6k}(x)\} = 5^{-1}S_{6k}$. Thus

$$U\{g^k(x)\} = \frac{F^{6k}(x)}{F^{6k}(x^5)}\frac{1}{x^k}\frac{1}{5}S_{6k} = \frac{S_{6k}}{5g^k(x)}.$$

(b) By the same calculation that we used in (a), we have

$$U\{\phi(x)g^k(x)\} = \frac{F^{6k}(x)}{F^{6k}(x^5)}\frac{1}{x^k}U\{\phi^{6k+1}(x)\}$$
$$= \frac{1}{g^k(x)}\frac{1}{5}S_{6k+1}.$$

(c) By (a), $U\{g^k(x)\} = S_{6k}/5g^k(x)$. By Lemma 4, $S_{6k} = \Sigma_{p=1}^{\infty} a_{6k,p}g^p$, where $a_{6k,p}$ is an integer divisible by 5 and $a_{6k,p} = 0$ unless $\left[\frac{6k+4}{5}\right] \leqslant p \leqslant 6k$. Clearly $\left[\frac{6k+4}{5}\right] \geqslant k$ for all nonnegative integers k and, in fact, $\left[\frac{6k+4}{5}\right] > k$ if $k \geqslant 1$. Thus the stated result holds for $U\{g^k(x)\}$.

By (b), $U\{\phi(x)g^k(x)\} = S_{6k+1}/5g^k(x)$. Again Lemma 4 implies that $S_{6k+1} = \Sigma_{p=1}^{\infty} a_{6k+1,p}g^p$, with $a_{6k+1,p}$ integers divisible by 5 and $a_{6k+1,p} = 0$ unless $[6k/5] + 1 \leqslant p \leqslant 6k + 1$. The result follows for $U\{\phi(x)g^k(x)\}$.

Theorem 6. For n a positive integer, $L_n(x)$ is a polynomial in g with integral coefficients divisible by 5 and no constant term. In particular, $L_1(x) = 5g(x)$.

Proof. By Lemma 5(b) with $k = 0$ we have $U\{\phi(x)\} = S_1/5$. On the other hand,

$$U\{\phi(x)\} = U\left\{F(x^{25})\frac{x}{F(x)}\right\} = F(x^5)U\{x/F(x)\}$$
$$= F(x^5)U\left\{\sum_{m=0}^{\infty} p(m)x^{m+1}\right\},$$

by Lemma 2(b) and the fact that

$$\frac{1}{F(x)} = \prod_{n=1}^{\infty} (1 - x^n)^{-1} = \sum_{m=0}^{\infty} p(m)x^m.$$

Applying U to the power series, we obtain

$$U\{\phi(x)\} = F(x^5) \sum_{m=0}^{\infty} p(5m + 4)x^{m+1} = L_1(x),$$

a result analogous to Lemma 3(b). Thus

$$L_1(x) = U\{\phi(x)\} = \frac{S_1}{5} = 5g,$$

by (12).

We complete the proof by induction on n. Suppose the theorem holds for $n = N \geqslant 1$. We consider L_{N+1}. Applying Lemma 3 we find that if N is odd, then $L_{N+1}(x) = U\{L_N(x)\}$, and if N is even, $L_{N+1}(x) = U\{\phi(x)L_N(x)\}$. In either case Lemma 5(c) and the linearity of U [Lemma 2(a)], together with the induction hypothesis, imply that $L_{N+1}(x)$ is a polynomial in g with integral coefficients divisible by 5 and no constant term. This completes the proof.

3. A DIGRESSION: THE RAMANUJAN IDENTITIES FOR POWERS OF THE PRIME 5

At this point we digress briefly from our principal goal, the proof of the Ramanujan congruences for powers of the prime 5, to discuss some related identities which were first discovered by Ramanujan [*Collected Papers*, (New York: Cambridge University Press, 1927), pp. 210–213] and later treated in an interesting way by Rademacher [*Trans. Am. Math. Soc., 51* (1942), pp. 609–636]. The identity originally stated by Ramanujan for the prime 5 is

$$\sum_{m=0}^{\infty} p(5m+4)x^m = 5\frac{\prod_{n=1}^{\infty}(1-x^{5n})^5}{\prod_{n=1}^{\infty}(1-x^n)^6}. \tag{14}$$

At the same time he gave a similar identity connected with the prime 7, but we postpone a discussion of this until Section 6.

We have already proved (14) in Theorem 6. For

$$L_1(x) = F(x^5)\sum_{m=0}^{\infty} p(5m+4)x^{m+1} \qquad \text{and} \qquad g(x) = xF^6(x^5)/F^6(x).$$

Putting these expressions into the equation $L_1(x) = 5g(x)$, we obtain

$$\sum_{m=0}^{\infty} p(5m+4)x^m = 5F^5(x^5)/F^6(x).$$

Replacing $F(x)$ by its infinite product representation yields (14). From (14) it follows that $\pi\{L_1(x)\} = \pi\{p(4)\} = 1$, so that by Proposition 1, the Ramanujan congruence modulo 5^1 is established. We can, in fact, use (14) to prove (2) for $n = 2$, that is, to prove the Ramanujan congruence modulo 5^2. For this purpose we need to recall Euler's identity and Jacobi's identity. These occurred as Corollary 5 and Corollary 6, respectively, in Section 2 of Chapter 3, and may be stated, respectively, as

$$F(x) = \sum_{m=-\infty}^{\infty} (-1)^m x^{m(3m+1)/2}$$

and

$$F(x)^3 = \sum_{m=0}^{\infty} (-1)^m(2m+1)x^{m(m+1)/2}.$$

By the binomial theorem $(1-x)^5 \equiv 1 - x^5 \pmod 5$, where the congruence is coefficient-wise. Thus, as a simple calculation shows, $(1-x)^5/(1-x^5) \equiv 1 \pmod 5$, and we conclude that $F^5(x)/F(x^5) \equiv 1 \pmod 5$. Hence (14) becomes

$$\begin{aligned} 5^{-1}\sum_{m=0}^{\infty} p(5m+4)x^m &= F^5(x^5)F^4(x)/F^{10}(x) \\ &\equiv F^5(x^5)F^4(x)/F^2(x^5) = F^3(x^5)F^4(x) \pmod 5. \end{aligned}$$

Using Euler's identity and Jacobi's identity, we have

$$5^{-1}\sum_{m=0}^{\infty} p(5m+4)x^m \equiv F^3(x^5)F^3(x)F(x)$$

$$(15)\qquad \equiv \sum_{j=0}^{\infty}(-1)^j(2j+1)x^{(5/2)j(j+1)} \times \sum_{k=0}^{\infty}(-1)^k(2k+1)x^{(1/2)k(k+1)} \times \sum_{l=-\infty}^{\infty}(-1)^l x^{(1/2)l(3l+1)} \pmod 5.$$

Now (2) with $n = 2$ states that $p(t) \equiv 0 \pmod{5^2}$, for all nonnegative integers $t \equiv 24 \pmod{5^2}$ and every such t has the form $t = 5m + 4$, with $m \equiv 4 \pmod 5$. On the other hand, the right-hand side of (15) has terms of the form x^m with $m = \frac{5}{2}j(j+1) + \frac{1}{2}k(k+1) + \frac{1}{2}l(3l+1)$, occurring with the coefficient $(-1)^{j+k+l}(2j+1)(2k+1)$. Clearly $m \equiv \frac{1}{2}k(k+1) + \frac{1}{2}l(3l+1)$ (mod 5), so that we are interested only in k,l such that $\frac{1}{2}k(k+1) + \frac{1}{2}l(3l+1) \equiv 4 \pmod 5$. As an elementary calculation shows, this happens only for $k \equiv 2$, $l \equiv 4 \pmod 5$. With this choice, however, the coefficient of $x^m \equiv (-1)^{j+k+l}(2j+1)(2k+1) \equiv 0 \pmod 5$. Thus follows from (15) that if $t \equiv 24 \pmod{5^2}$, then $5^{-1}p(t) \equiv 0 \pmod 5$, or $p(t) \equiv 0 \pmod{5^2}$, as was to be proved.

It should be clear at this point that Theorem 6 can be regarded as a generalization of (14), valid for all powers of 5. As an example we shall write down the explicit form of Theorem 6 for the case $n = 2$. An elementary (though tedious) calculation using (12) and (13) yields

$$S_6 = 5^{12}g^6 + 6\cdot 5^{10}g^5 + 63\cdot 5^7 g^4 + 52\cdot 5^2 g^3 + 63\cdot 5^2 g^2.$$

$L_2(x) = U\{L_1(x)\} = 5U\{g(x)\} = S_6/g$, so that, in fact,

$$L_2(x) = 5^{12}g^5 + 6\cdot 5^{10}g^4 + 63\cdot 5^7 g^3 + 52\cdot 5^5 g^2 + 63\cdot 5^2 g.$$

Since $L_2(x) = F(x)\Sigma_{m=0}^{\infty}p(5^2m+24)x^{m+1}$ and $g(x) = xF^6(x^5)/F^6(x)$, this equation takes the final form

$$(16)\qquad \sum_{m=0}^{\infty} p(5^2m+24)x^m = 5^{12}x^4\frac{F^{30}(x^5)}{F^{31}(x)} + 6\cdot 5^{10}x^3\frac{F^{24}(x^5)}{F^{25}(x)} + 63\cdot 5^7x^2\frac{F^{18}(x^5)}{F^{19}(x)} + 52\cdot 5^5 x\frac{F^{12}x^5}{F^{13}(x)} + 63\cdot 5^2\frac{F^6(x^5)}{F^7(x)}.$$

4. COMPLETION OF THE PROOF FOR POWERS OF 5

In this section we complete the proof of the Ramanujan congruences for powers of 5 by proving equation (5) in a revised form. The revision is contained in

Theorem 7. In view of Theorem 6, put

$$L_{2n-1}(x) = \sum_{s=1}^{\infty} b_{n,s} g^s(x), \qquad L_{2n}(x) = \sum_{s=1}^{\infty} c_{n,s} g^s(x),$$

where $b_{n,s}$ and $c_{n,s}$ are integers divisible by 5, and for each n, $b_{n,s}$ and $c_{n,s}$ are zero for sufficiently large s (depending on n, of course). Then equation (5) of Proposition 1 is equivalent to

$$\begin{aligned} &\pi(b_{n,1}) = 2n - 1, \qquad &\pi(c_{n,1}) = 2n, \\ &\pi(b_{n,s}) \geqslant 2n - 1, \qquad &\pi(c_{n,s}) \geqslant 2n, \end{aligned} \tag{17}$$

for each positive integer n.

Proof. We begin by proving

$$p(l_{2n-1}) = b_{n,1}, \qquad p(l_{2n}) = c_{n,1}. \tag{18}$$

From the definition of $L_{2n-1}(x)$ it is clear that $p(l_{2n-1})$ is the coefficient of x in the power series for $L_{2n-1}(x)$. On the other hand, $g^s(x) = x^s +$ higher powers of x. Thus for $s > 1$, $g^s(x)$ has no term of the form x, and, comparing coefficients of x in $L_{2n-1}(x) = \Sigma_{s=1}^{\infty} b_{n,s} g^s(x)$, we conclude that $p(l_{2n-1}) = b_{n,1}$. The same argument applied to the equation $L_{2n}(x) = \Sigma_{s=1}^{\infty} c_{n,s} g^s(x)$ yields $p(l_{2n}) = c_{n,1}$. Thus (18) is proved.

Since $g(x)$ can be expressed as a power series with integral coefficients, it is clear from (18) and the definition of $b_{n,s}$ and $c_{n,s}$ that (17) implies (5). To prove the converse, suppose that (5) holds. We derive (17) by induction on s. By (18) and (5), $\pi(b_{n,1}) = \pi\{p(l_{2n-1})\} = 2n - 1$, and $\pi(c_{n,1}) = \pi\{p(l_{2n})\} = 2n$, so that (17) holds for $s = 1$. Suppose $s_0 \geqslant 1$ and (17) holds for all $s \leqslant s_0$. Since $g(x) = x\{1 + P(x)\}$, where $P(x)$ is a power series in x with integral coefficients, we find that the coefficient d_{s_0+1} of x^{s_0+1} in $L_{2n-1}(x)$ has the form $b_{n,s_0+1} + \alpha_1 b_{n,s_0} + \alpha_2 b_{n,s_0-1} + \cdots + \alpha_{s_0} b_{n,1}$, where the α_i are integers depending on n. By the induction hypothesis, $\pi(\alpha_1 b_{n,s_0} + \cdots + \alpha_{s_0} b_{n,1}) \geqslant 2n - 1$. Thus if $\pi(b_{n,s_0+1}) < 2n - 1$, then (4) would imply that $\pi(d_{s_0+1}) < 2n - 1$, in contradiction to (5). Thus $\pi(b_{n,s}) \geqslant 2n - 1$ for all $s \geqslant 1$. In the same way, $\pi(c_{n,s}) \geqslant 2n$ for all $s \geqslant 1$.

We complete the proof of the Ramanujan congruences modulo powers of 5 by proving the following theorem, which is actually somewhat stronger than (17).

Theorem 8. For each positive integer n we have

$$\pi(b_{n,1}) = 2n - 1, \qquad \pi(b_{n,s}) \geqslant 2n - 1 + \left[\frac{5s - 5}{2}\right] \tag{19}$$

and

$$\pi(c_{n,1}) = 2n, \qquad \pi(c_{n,s}) \geqslant 2n + \left[\frac{5s - 4}{2}\right]. \tag{20}$$

Proof. The proof is by induction on n and has three basic steps:
(a) proof that (19) holds for $n = 1$,
(b) proof that if (19) holds for n, then (20) also holds for n,
(c) proof that if (20) holds for n, then (19) holds for $n + 1$.

(a) Let $n = 1$. We proved that $L_1(x) = 5g(x)$. Thus $b_{1,1} = 5$ and $b_{1,s} = 0$ for all $s > 1$. Obviously, then, $\pi(b_{1,1}) = 1$ and $\pi(b_{1,s}) \geqslant 1 + \left[\frac{5s - 5}{2}\right]$.

(b) By Lemmas 3(a) and 5(a),

$$\sum_{s=1}^{\infty} c_{n,s}g^s(x) = L_{2n}(x) = U\{L_{2n-1}(x)\} = U\left\{\sum_{\sigma=1}^{\infty} b_{n,\sigma}g^{\sigma}(x)\right\}$$

$$= \sum_{\sigma=1}^{\infty} b_{n,\sigma}U\{g^{\sigma}(x)\} = \sum_{\sigma=1}^{\infty} b_{n,\sigma}\frac{S_{6\sigma}}{5g^{\sigma}(x)}.$$

By Lemma 4, $S_{6\sigma} = \Sigma_{p=1}^{\infty} a_{6\sigma,p}g^p$, with $\pi(a_{6\sigma,p}) \geqslant \left[\frac{5p - 6\sigma + 1}{2}\right]$. It follows that $5c_{n,s} = \Sigma_{\sigma=1}^{\infty} b_{n,\sigma}a_{6\sigma,s+\sigma}$, a finite sum, and thus

$$1 + \pi(c_{n,s}) = \pi(5c_{n,s}) = \pi\left(\sum_{\sigma=1}^{\infty} b_{n,\sigma}a_{6\sigma,s+\sigma}\right)$$

$$\geqslant \min_{\sigma \geqslant 1} \{\pi(b_{n,\sigma}) + \pi(a_{6\sigma,s+\sigma})\},$$

by (4).

We now assume that (19) holds for n. Thus

$$1 + \pi(c_{n,s}) \geqslant \min_{\sigma \geqslant 1} \left\{2n - 1 + \left[\frac{5\sigma - 5}{2}\right] + \left[\frac{5s - \sigma + 1}{2}\right]\right\},$$

or

$$\pi(c_{n,s}) \geqslant 2n - 2 + \min_{\sigma \geqslant 1} \left\{\left[\frac{5\sigma - 5}{2}\right] + \left[\frac{5s - \sigma + 1}{2}\right]\right\}.$$

However, if we increase σ by 1, this increases $\left[\frac{5\sigma - 5}{2}\right]$ by $\geqslant 2$ and decreases $\left[\frac{5s - \sigma + 1}{2}\right]$ by $\leqslant 1$. Hence $\left[\frac{5\sigma - 5}{2}\right] + \left[\frac{5s - \sigma + 1}{2}\right]$, with fixed s, is a strictly monotone increasing function of $\sigma \geqslant 1$. Thus the above minimum is attained for $\sigma = 1$ and we have $\pi(c_{n,s}) \geqslant 2n - 2 + \left[\frac{5s}{2}\right] = 2n + \left[\frac{5s - 4}{2}\right]$. This is the second part of (20) for n.

If $s = 1$ we obtain $\pi(c_{n,1}) = -1 + \pi(\sum_{\sigma=1}^{\infty} b_{n,\sigma} a_{6\sigma,1+\sigma})$. We need to have $\pi(a_{6,2})$ explicitly. Since $a_{6,2}$ is the coefficient of g^2 in S_6, it follows from Section 3 that $a_{6,2} = 63 \cdot 5^2$, and thus $\pi(a_{6,2}) = 2$. We conclude that $\pi(b_{n,1} a_{6,2}) = \pi(b_{n,1}) + \pi(a_{6,2}) = 2n - 1 + 2 = 2n + 1$, by (19). On the other hand, by (4) and (19),

$$\pi\left(\sum_{\sigma=2}^{\infty} b_{n,\sigma} a_{6\sigma,1+\sigma}\right) \geqslant \min_{\sigma \geqslant 2} \{\pi(b_{n,\sigma}) + \pi(a_{6\sigma,1+\sigma})\}$$

$$\geqslant 2n - 1 + \min_{\sigma \geqslant 2} \left\{\left[\frac{5\sigma - 5}{2}\right] + \left[\frac{6 - \sigma}{2}\right]\right\}.$$

As before, $\left[\frac{5\sigma - 5}{2}\right] + \left[\frac{6 - \sigma}{2}\right]$ is a strictly monotone increasing function of σ, so that the minimum is attained for $\sigma = 2$, and

$$\pi\left(\sum_{\sigma=2}^{\infty} b_{n,\sigma} a_{6\sigma,1+\sigma}\right) \geqslant 2n - 1 + 2 + 2 = 2n + 3 > 2n + 1.$$

It follows from (4) that

$$\pi\left(\sum_{\sigma=1}^{\infty} b_{n,\sigma} a_{6\sigma,1+\sigma}\right) = 2n + 1,$$

and we conclude that $\pi(c_{n1}) = 2n$. Hence (20) holds for n.

(c) The proof here is similar to that of (b). Assume that (20) holds for n. By Lemmas 3(b) and 5(b),

$$\sum_{s=1}^{\infty} b_{n+1,s} g^s(x) = L_{2n+1}(x) = U\{\phi(x) L_{2n}(x)\}$$

$$= U\left\{\phi(x) \sum_{\sigma=1}^{\infty} c_{n,\sigma} g^{\sigma}(x)\right\} = \sum_{\sigma=1}^{\infty} c_{n,\sigma} U\{\phi(x) g^{\sigma}(x)\}$$

$$= \sum_{\sigma=1}^{\infty} c_{n,\sigma} \frac{S_{6\sigma+1}}{5 g^{\sigma}(x)} = \sum_{\sigma=1}^{\infty} \frac{c_{n,\sigma}}{5} \sum_{p=1}^{\infty} a_{6\sigma+1,p} g^{p-\sigma}.$$

It follows that $5b_{n+1,s} = \Sigma_{\sigma=1}^{\infty} c_{n,\sigma}a_{6\sigma+1,\sigma+s}$, a finite sum. Thus, by (4),

$$1 + \pi(b_{n+1,s}) = \pi(5b_{n+1,s}) \geqslant \min_{\sigma\geqslant 1} \{\pi(c_{n,\sigma}) + \pi(a_{6\sigma+1,\sigma+s})\}.$$

By Lemma 4, $\pi(a_{6\sigma+1,\sigma+s}) \geqslant \left[\frac{5s-\sigma}{2}\right]$, so that

$$\pi(b_{n+1,s}) \geqslant 2n - 1 + \min_{\sigma\geqslant 1}\left\{\left[\frac{5\sigma-4}{2}\right] + \left[\frac{5s-\sigma}{2}\right]\right\}.$$

But $\left[\frac{5\sigma-4}{2}\right] + \left[\frac{5s-\sigma}{2}\right]$ is a strictly monotone increasing function of σ for $\sigma \geqslant 1$. Hence the minimum occurs for $\sigma = 1$ and we have

$$\pi(b_{n+1,s}) \geqslant 2n - 1 + \left[\frac{5s-1}{2}\right] = 2n + 1 + \left[\frac{5s-5}{2}\right],$$

the second part of (19) for $n + 1$.

With $s = 1$, we have

$$\pi(b_{n+1,1}) = -1 + \pi\left(c_{n,1}a_{7,2} + \sum_{\sigma=2}^{\infty} c_{n,\sigma}a_{6\sigma+1,\sigma+1}\right).$$

Now $\pi(c_{n,1}) = 2n$, by (20), and a calculation of S_7 yields $a_{7,2} = 28 \cdot 5^2$, so that $\pi(a_{7,2}) = 2$. Hence $\pi(c_{n,1}a_{7,2}) = 2n + 2$, by (4). On the other hand, by (4) and (20),

$$\pi\left(\sum_{\sigma=2}^{\infty} c_{n,\sigma}a_{6\sigma+1,\sigma+1}\right) \geqslant \min_{\sigma\geqslant 2} \{\pi(c_{n,\sigma}) + \pi(a_{6\sigma+1,\sigma+1})\}$$

$$\geqslant 2n + \min_{\sigma\geqslant 2}\left\{\left[\frac{5\sigma-4}{2}\right] + \left[\frac{5-\sigma}{2}\right]\right\}.$$

Once more the function that we are minimizing is a strictly monotone increasing function of σ. Thus the minimum occurs for $\sigma = 2$, and

$$\pi\left(\sum_{\sigma=2}^{\infty} c_{n,\sigma}a_{6\sigma+1,\sigma+1}\right) \geqslant 2n + 3 + 1 = 2n + 4 > 2n + 2.$$

By (4) we conclude that $\pi(b_{n+1,1}) = 2n + 1$, and the proof is complete.

5. START OF THE PROOF FOR POWERS OF 7

The broad outline of the proof of the Ramanujan congruences for powers of 7 is the same as that for powers of 5. Thus we shall give the argument in a sketchier form in this case, avoiding, when possible, tedious details that are merely a repetition of those already given in the case of the prime 5.

For this case we define

(21) $$l_{2n-1} = (17 \cdot 7^{2n-1} + 1)/24, \qquad l_{2n} = (23 \cdot 7^{2n} + 1)/24,$$

with n a positive integer, and it turns out that l_n is the smallest positive integral solution of $24x \equiv 1 \pmod{7^n}$. Thus (1b) of Chapter 7 can be rewritten as

(22) $$p(7^n m + l_n) \equiv 0 \pmod{7^{[(n+2)/2]}}, \qquad m \geqslant 0,$$

for each positive integer n. In addition, we shall show that the power of 7 occurring in (22) is the best possible in the sense that

(23) $$p(l_n) \not\equiv 0(7^{[(n+2)/2]+1}),$$

for each positive integer n.

We now introduce the following functions, defined for $|x| < 1$:

$$\phi(x) = x^2 F(x^{49})/F(x)$$

$$L_{2n-1}(x) = F(x^7) \sum_{m=0}^{\infty} p(7^{2n-1}m + l_{2n-1})x^{m+1}$$

$$L_n(x) = F(x) \sum_{m=0}^{\infty} p(7^{2n}m + l_{2n})x^{m+1},$$

where n is a positive integer and $F(x)$ is the function already defined in Section 1 of this chapter. In this case $\phi(e^{2\pi i\tau}) = \eta(49\tau)/\eta(\tau) = h_7(\tau) = h(\tau)$, for $\tau \in \mathscr{H}$. We also define $g(x) = xF^4(x^7)/F^4(x)$, for $|x| < 1$. It is clear that for $\tau \in \mathscr{H}$, $g(e^{2\pi i\tau}) = \Phi_{7,4}(\tau)$. Furthermore, we define the integer-valued function $\pi\{f(x)\}$ as it was defined in Section 1, except that the prime 7 replaces the prime 5. The condition (4) is still fulfilled. Finally we define the linear operator $U_7 = U$ by†

$$U\{f(x)\} = \sum_{7n \geqslant N} \lambda_{7n} x^n,$$

when $f(x) = \sum_{n=N}^{\infty} \lambda_n x^n$. With these definitions the calculations of Section 1 yield the following results.

Lemma 9. (a) If w is a primitive seventh root of unity, then $7U\{f(x)\} = \sum_{r=0}^{\infty} f(w^r x^{1/7})$.

(b) For n a positive integer, $U\{L_{2n-1}(x)\} = L_{2n}(x)$ and $U\{\phi(x)L_{2n}(x)\} = L_{2n+1}(x)$.

(c) $U\{f_2(x^7)f_1(x)\} = f_2(x)U\{f_1(x)\}$.

† U_7 is a Hecke operator.

Proposition 10. For each positive integer n, the equation

$$\pi\{L_n(x)\} = \pi\{p(l_n)\} = [(n+2)/2] \tag{24}$$

is equivalent to (22) and (23) together.

We now apply the modular equation for the prime 7 that we derived in Section 7 of Chapter 7 in the form

$$\begin{aligned}\Phi(\tau)^2 = {} & \Phi(\tau)\{7^2h(\tau)^3 + 7\cdot 5h(\tau)^2 + 7h(\tau)\}\\ & + \{7^3h(\tau)^7 + 7^3h(\tau)^6 + 3\cdot 7^2h(\tau)^5 + 7^2h(\tau)^4 + 3\cdot 7h(\tau)^3\\ & + 7h(\tau)^2 + h(\tau)\},\end{aligned}$$

where $\Phi(\tau) = \Phi_{7,4}(\tau) = \{\eta(7\tau)/\eta(\tau)\}^4$ and $h(\tau) = h_7(\tau) = \eta(49\tau)/\eta(\tau)$. We rearrange this in an elementary way to obtain

$$\begin{aligned}h(\tau)^7 & - \Phi(7\tau)\{7^2h^6(\tau) + 7\cdot 5h^5(\tau) + 7h^4(\tau)\}\\ & - \Phi^2(7\tau)\{7^3h^6(\tau) + 7^3h^5(\tau) + 3\cdot 7^2h^4(\tau) + 7^2h^3(\tau) + 3\cdot 7h^2(\tau)\\ & + 7h(\tau) + 1\} = 0,\end{aligned}$$

or

$$\begin{aligned}&\phi(x)^7 - g(x^7)\{7^2\phi(x)^6 + 7\cdot 5\phi(x)^5 + 7\phi(x)^4\}\\ &-g^2(x^7)\{7^3\phi(x)^6 + 7^3\phi(x)^5 + 3\cdot 7^2\phi(x)^4 + 7^2\phi(x)^3 + 3\cdot 7\phi(x)^2\\ &+7\phi(x) + 1\} = 0.\end{aligned} \tag{25}$$

From (25) it follows that the seven distinct roots of the polynomial

$$\begin{aligned}&u^7 - g(x)\{7^2u^6 + 7\cdot 5u^5 + 7u^4\}\\ &- g^2(x)\{7^3u^6 + 7^3u^5 + 3\cdot 7^2u^4 + 7^2u^3 + 3\cdot 7u^2 + 7u + 1\}\end{aligned} \tag{26}$$

are $\phi(w^kx^{1/7})$, where w is a primitive seventh root of unity and $0 \leqslant k \leqslant 6$. To apply Newton's formula put $S_r = \Sigma_{k=0}^{6}\, \phi^r(w^kx^{1/7})$. By Lemma 9(a), of course, $S_r = 7U\{\phi^r(x)\}$, a fact which we shall apply later. An application of Newton's formula to the polynomial (26) yields

Lemma 11. Let r be a positive integer. Then S_r is a polynomial in g of the form $S_r = \Sigma_{p=1}^{\infty}\, a_{rp}g^p$, where a_{rp} is an integer divisible by 7, such that $\pi(a_{rp}) \geqslant \left[\dfrac{7p-2r+3}{4}\right]$ and $a_{rp} = 0$ unless $\left[\dfrac{2r+6}{7}\right] \leqslant p \leqslant 2r$.

Proof. By Newton's formula we have

$$\begin{aligned}S_1 &= 7^3g^2 + 7^2g,\\ S_2 &= 7^6g^4 + 2\cdot 7^5g^3 + 9\cdot 7^3g^2 + 10\cdot 7g,\\ S_3 &= 7^9g^6 + 3\cdot 7^8g^5 + 24\cdot 7^6g^4 + 85\cdot 7^4g^3 + 114\cdot 7^2g^2 + 3\cdot 7g,\end{aligned} \tag{27}$$

and, more generally,

$$(28)\qquad S_r = \begin{cases} c_1S_{r-1} - c_2S_{r-2} + \cdots + (-1)^r c_{r-1}S_1 + (-1)^{r+1}c_r r & \text{if } 1 \leqslant r \leqslant 7, \\ c_1S_{r-1} - c_2S_{r-2} + \cdots + c_7S_{r-7} & \text{if } r > 7, \end{cases}$$

where $c_1 = 7^2g + 7^3g^2$, $c_2 = -(7\cdot 5g + 7^3g^2)$, $c_3 = 7g + 3\cdot 7^2g^2$, $c_4 = -7^2g^2$, $c_5 = 3\cdot 7g^2$, $c_6 = -7g^2$, $c_7 = g^2$.

By (27) it is clear that the lemma holds for $r = 1, 2, 3$. The proof can be completed by induction on r, with (28) playing a key role. We carry out the details only for the inequality on $\pi(a_{rp})$. Assume the inequality holds for all $r < R$. If $2 \leqslant R \leqslant 7$, then by (28),

$$a_{R,p} = 7^2a_{R-1,p-1} + 7^3a_{R-1,p-2} + 7\cdot 5\,a_{R-2,p-1} + 7^3a_{R-2,p-2} + \cdots.$$

By (4) it follows that

$$\begin{aligned}\pi(a_{R,p}) &\geqslant \min\left\{2 + \left[\frac{7p-2R-2}{4}\right], 3 + \left[\frac{7p-2R-9}{4}\right],\right.\\ &\qquad \left. 1 + \left[\frac{7p-2R}{4}\right], 3 + \left[\frac{7p-2R-7}{4}\right], \ldots\right\}\\ &= \left[\frac{7p-2R+3}{4}\right],\end{aligned}$$

which is the desired inequality. If $R > 7$, then by (28),

$$\begin{aligned}a_{R,p} &= 7^2a_{R-1,p-1} + 7^3a_{R-1,p-2} + 7\cdot 5a_{R-2,p-1} + 7^3a_{R-2,p-2}\\ &\quad + 7a_{R-3,p-1} + 3\cdot 7^2a_{R-3,p-2} + 7^2a_{R-4,p-2}\\ &\quad + 3\cdot 7a_{R-5,p-2} + 7a_{R-6,p-2} + a_{R-7,p-2}.\end{aligned}$$

By (4) we have

$$\begin{aligned}\pi(a_{R,p}) &\geqslant \min\left\{\left[\frac{7p-2R+6}{4}\right], \left[\frac{7p-2R+3}{4}\right],\right.\\ &\qquad \left.\left[\frac{7p-2R+4}{4}\right], \left[\frac{7p-2R+5}{4}\right]\right\}\\ &= \left[\frac{7p-2R+3}{4}\right],\end{aligned}$$

the required inequality, and the lemma follows by induction.

Lemma 12. If k is a nonnegative integer, then we have the following:

(a) $U\{g^k(x)\} = S_{4k}/7g^k(x)$.

(b) $U\{\phi(x)g^k(x)\} = S_{4k+1}/7g^k(x)$.

(c) Both $U\{g^k(x)\}$ and $U\{\phi(x)g^k(x)\}$ are polynomials in $g(x)$ with integral coefficients. If $k \geqslant 1$, the polynomial has no constant term.

Proof. (a)

$$U\{g^k(x)\} = U\{x^k F^{4k}(x^7)/F^{4k}(x)\} = U\left\{\frac{F^{4k}(x^7)}{F^{4k}(x^{49})}\frac{\phi^{4k}(x)}{x^{7k}}\right\}$$

$$= \frac{F^{4k}(x)}{F^{4k}(x^7)}\frac{1}{x^k}U\{\phi^{4k}(x)\} = \frac{1}{g^k(x)}\frac{S_{4k}}{7},$$

by Lemma 9(c) and our previous calculation of $U\{\phi^{4k}(x)\}$.

(b)
$$U\{\phi(x)g^k(x)\} = \frac{1}{g^k(x)}U\{\phi^{4k+1}(x)\} = \frac{S_{4k+1}}{7g^k(x)},$$

by the same calculation as in (a).

(c) By (a), $U\{g^k(x)\} = S_{4k}/7g^k(x)$. By Lemma 11, $S_{4k} = \Sigma_{p=1}^{\infty} a_{4k,p}g^p$, where $a_{4k,p}$ is an integer divisible by 7 and $a_{4k,p} = 0$ unless $\left\lceil\frac{8k+6}{7}\right\rceil \leqslant p \leqslant 8k$. We have $\left\lceil\frac{8k+6}{7}\right\rceil \geqslant k$ for all nonnegative integers k and $\left\lceil\frac{8k+6}{7}\right\rceil > k$ if $k \geqslant 1$. The stated result then follows for $U\{g^k(x)\}$. The proof for $U\{\phi(x)g^k(x)\}$ is quite similar.

Theorem 13. For any positive integer n, $L_n(x)$ is a polynomial in g with integral coefficients divisible by 7, and no constant term. In particular, $L_1(x) = 7g(x) + 7^2g^2(x)$.

Proof. By Lemma 12(b) with $k = 0$ we have $U\{\phi(x)\} = S_1/7$. On the other hand,

$$U\{\phi(x)\} = U\left\{F(x^{49})\frac{x^2}{F(x)}\right\} = F(x^7)U\{x^2/F(x)\}$$

$$= F(x^7)U\left\{\sum_{m=0}^{\infty} p(m)x^{m+2}\right\},$$

by Lemma 9(c). We thus obtain

$$U\{\phi(x)\} = F(x^7)\sum_{m=0}^{\infty} p(7m+5)x^{m+1} = L_1(x),$$

after applying the definition of U_7 to the power series. Hence

$$L_1(x) = U\{\phi(x)\} = S_1/7 = 7^2g^2(x) + 7g(x),$$

by (27).

Suppose the result holds for $n = N \geqslant 1$. By Lemma 9(b), $L_{N+1}(x) = U\{L_N(x)\}$, if N is odd, and $L_{N+1}(x) = U\{\phi(x)L_N(x)\}$, if N is even. In either case Lemma 12(c) and the linearity of U imply that $L_{N+1}(x)$ is a polynomial in g with integral coefficients divisible by 7, and no constant term. The theorem follows by induction.

6. A SECOND DIGRESSION: THE RAMANUJAN IDENTITIES FOR POWERS OF THE PRIME 7

Here we discuss the Ramanujan identity connected with the prime 7, that was mentioned in Section 3. The identity in question is

$$\sum_{m=0}^{\infty} p(7m+5)x^m = 7\frac{\prod_{n=1}^{\infty}(1-x^{7n})^3}{\prod_{n=1}^{\infty}(1-x^n)^5} + 7^2x\frac{\prod_{n=1}^{\infty}(1-x^{7n})^7}{\prod_{n=1}^{\infty}(1-x^n)^8}. \tag{29}$$

The identity (29) is already contained in Theorem 13, for

$$L_1(x) = F(x^7)\sum_{m=0}^{\infty} p(7m+5)x^{m+1}$$

and $g(x) = xF^4(x^7)/F^4(x)$. Putting these expressions into the equation $L_1(x) = 7g(x) + 7^2g^2(x)$, we obtain (29). From (29) it follows that $\pi\{L_1(x)\} = \pi\{p(5)\} = 1$. Hence by Proposition 10 the Ramanujan congruence modulo 7^1 follows.

The Ramanujan congruence modulo 7^2 states that $p(7^2m + 47) \equiv 0 \pmod{7^2}$ for $m \geqslant 0$. We shall now apply (29) to obtain this congruence and somewhat more. We, in fact, shall prove, using (29), that for all nonnegative integers m,

$$p(7^2m+19) \equiv p(7^2m+33) \equiv p(7^2m+40) \equiv p(7^2m+47) \equiv 0 \pmod{7^2}.$$

These congruences were given by Zuckerman [*Duke Math J.* 5 (1939), pp. 88–110, esp. p. 89]. To prove them we first write (29) in the form

$$\frac{1}{7}\sum_{m=0}^{\infty} p(7m+5)x^m = \frac{F(x^7)^3}{F(x)^4} + 7x\frac{F(x^7)^7}{F(x)^8}.$$

Using the binomial theorem, we find that $(1-x)^7/(1-x^7) \equiv 1 \pmod 7$, and as a consequence, $F^7(x)/F(x^7) \equiv 1 \pmod 7$. Thus

$$\frac{1}{7}\sum_{m=0}^{\infty} p(7m+5)x^m \equiv \frac{F(x^7)^3}{F(x)^4} \equiv \frac{F(x^7)^3F(x)^3}{F(x)^7} \equiv F(x^7)^2F(x)^3 \pmod 7.$$

Applying the Euler and Jacobi identities yields

$$(30)\qquad \begin{aligned}\frac{1}{7}\sum_{m=0}^{\infty} p(7m+5)x^m \equiv \sum_{j=-\infty}^{\infty} (-1)^j x^{(7/2)j(3j+1)} \sum_{k=-\infty}^{\infty} (-1)^k x^{(7/2)k(3k+1)} \\ \times \sum_{l=0}^{\infty} (-1)^l (2l+1) x^{(1/2)l(l+1)} \pmod 7.\end{aligned}$$

The congruences to be proved can be stated in the form

$$(31)\qquad p(t) \equiv 0 \pmod{7^2}, \qquad \text{for } t \equiv 19, 33, 40, \text{ or } 47 \pmod{7^2}.$$

All t occurring in (31) have the form $t = 7m + 5$, with m a nonnegative integer. Furthermore, with this representation, $t \equiv 19, 33, 40$, or $47 \pmod{7^2}$, according as $m \equiv 2, 4, 5$, or $6 \pmod 7$. The right-hand side of (30) has terms of the form x^m with $m = \frac{7}{2}j(3j+1) + \frac{7}{2}k(3k+1) + \frac{1}{2}l(l+1)$, occurring with the coefficient $(-1)^{j+k+l}(2l+1)$. Clearly the m which occur are $\equiv \frac{1}{2}l(l+1) \pmod 7$, so we want to consider only l such that $\frac{1}{2}l(l+1) \equiv 2, 4, 5$, or $6 \pmod 7$. A simple calculation shows that $\frac{1}{2}l(l+1)$ is never $\equiv 2$, 4, or 5 (mod 7), and $\equiv 6 \pmod 7$ only if $l \equiv 3 \pmod 7$. With $l \equiv 3 \pmod 7$ the coefficient $(-1)^{j+k+l}(2l+1)$ is $\equiv 0 \pmod 7$. It follows from (30) that if $m = 2, 4, 5$, or $6 \pmod 7$, then $\frac{1}{7}p(7m+5) \equiv 0 \pmod 7$, or equivalently, $p(7m+5) \equiv 0 \pmod{7^2}$. Thus (31) follows and the proof is complete.

Theorem 13 can be regarded as a generalization of (29), valid for all powers of 7. The identity for 7^2 corresponding to (29) is derived from the following calculation:

$$\begin{aligned} L_2(x) &= U\{L_1(x)\} = U\{7g(x) + 7^2 g^2(x)\} \\ &= 7U\{g(x)\} + 7^2 U\{g^2(x)\} = \frac{S_4}{g(x)} + 7\frac{S_8}{g^2(x)}. \end{aligned}$$

Unfortunately the calculation of S_8 in terms of $g(x)$ without a computing machine would be a most tedious affair, and we do not carry it out here.

7. COMPLETION OF THE PROOF FOR POWERS OF 7

We resume the proof of the Ramanujan congruences for powers of 7 with **Theorem 14.** As a consequence of Theorem 13 we can write

$$L_{2n-1}(x) = \sum_{s=1}^{\infty} b_{n,s} g^s(x), \qquad L_{2n}(x) = \sum_{s=1}^{\infty} c_{n,s} g^s(x),$$

where $b_{n,s}$ and $c_{n,s}$ are integers divisible by 7, and for each n, $b_{n,s}$ and $c_{n,s}$ are zero for sufficiently large s, depending on n. Then (24) of Proposition 10 is equivalent to

$$\begin{aligned} &\pi(b_{n,1}) = n, \qquad \pi(c_{n,1}) = n + 1, \\ &\pi(b_{n,s}) \geqslant n, \qquad \pi(c_{n,s}) \geqslant n + 1. \end{aligned} \tag{32}$$

Proof. We claim that

$$p(l_{2n-1}) = b_{n,1}, \qquad p(l_{2n}) = c_{n,1}. \tag{33}$$

Clearly $p(l_{2n-1})$ is the coefficient of x in the series for $L_{2n-1}(x)$ and $p(l_{2n})$ is the coefficient of x in the series for $L_{2n}(x)$. Since the series for $g(x)$ has the form $x +$ higher powers of x, (33) follows immediately from a comparison of the coefficients of x in

$$L_{2n-1}(x) = \sum_{s=1}^{\infty} b_{n,s} g^s(x) \qquad \text{and} \qquad L_{2n}(x) = \sum_{s=1}^{\infty} c_{n,s} g^s(x).$$

Since $g(x)$ is a power series in x with integral coefficients and first coefficient 1, and since $\left[\dfrac{2n+2}{2}\right] = n + 1$, $\left[\dfrac{2n+1}{2}\right] = n$, (33) implies that (24) is equivalent to (32), as in the proof of Theorem 7.

We complete the proof of the Ramanujan congruences for powers of 7 by proving the following result, which clearly implies (32).

Theorem 15. For each positive integer n we have

$$\pi(b_{n,1}) = n, \qquad \pi(b_{n,s}) \geqslant n + \left[\frac{7s-4}{4}\right], \tag{34}$$

with the exception $\pi(b_{1,2}) = 2$, and

$$\pi(c_{n,1}) = n + 1, \qquad \pi(c_{n,s}) \geqslant n + 1 + \left[\frac{7s-6}{4}\right]. \tag{35}$$

Proof. Like the proof of Theorem 8, the proof of this theorem is by induction on n and has three basic steps:

(a) We first show that (34) holds with $n = 1$, but taking into account the fact that this is the exceptional case of (34). In Theorem 13 we proved that $L_1(x) = 7g(x) + 7^2 g^2(x)$. Thus $b_{1,1} = 7$, $b_{1,2} = 7^2$, and $b_{1,s} = 0$ for all $s > 2$. Hence (34) holds for $n = 1$ with the exceptional value $\pi(b_{1,2}) = 2$.

(b) We next show that if (34) holds for a fixed $n \geqslant 2$, then (35) also holds for this value of n. Because of the exceptional case of (34) for $n = 1$, we prove separately that (35) holds for $n = 1$. To begin, suppose $n \geqslant 1$. By Lemmas

9(b) and 12(a),

$$\sum_{s=1}^{\infty} c_{n,s} g^s(x) = L_{2n}(x) = U\{L_{2n-1}(x)\} = U\left\{\sum_{\sigma=1}^{\infty} b_{n,\sigma} g^\sigma(x)\right\}$$

$$= \sum_{\sigma=1}^{\infty} b_{n,\sigma} U\{g^\sigma(x)\} = \sum_{\sigma=1}^{\infty} b_{n,\sigma} \frac{S_{4\sigma}}{7g^\sigma(x)}.$$

By Lemma 11, $S_{4\sigma} = \Sigma_{p=1}^{\infty} a_{4\sigma,p} g^p$, a finite sum, with $\pi(a_{4\sigma,p}) \geqslant \left[\frac{7p - 8\sigma + 3}{4}\right]$. It follows that $7c_{n,s} = \Sigma_{\sigma=1}^{\infty} b_{n,\sigma} a_{4\sigma,s+\sigma}$, a finite sum, and we conclude that

$$1 + \pi(c_{n,s}) = \pi(7c_{n,s}) \geqslant \min_{\sigma \geqslant 1} \{\pi(b_{n,\sigma}) + \pi(a_{4\sigma,s+\sigma})\},$$

by (4). Now assume that (34) holds with a fixed $n \geqslant 2$. Then

$$1 + \pi(c_{n,s}) \geqslant \min_{\sigma \geqslant 1} \left\{n + \left[\frac{7\sigma - 4}{4}\right] + \left[\frac{7s - \sigma + 3}{4}\right]\right\},$$

or

$$\pi(c_{n,s}) \geqslant n - 1 + \min_{\sigma \geqslant 1} \left\{\left[\frac{7\sigma - 4}{4}\right] + \left[\frac{7s - \sigma + 3}{4}\right]\right\}.$$

If we increase σ by 1, then $\left[\frac{7\sigma - 4}{4}\right]$ increases by $\geqslant 1$, while $\left[\frac{7s - \sigma + 3}{4}\right]$ decreases by $\leqslant 1$. Thus with fixed s, $\left[\frac{7\sigma - 4}{4}\right] + \left[\frac{7s - \sigma + 3}{4}\right]$ is a monotone nondecreasing function of $\sigma \geqslant 1$, and the minimum is attained for $\sigma = 1$. We conclude, therefore, that $\pi(c_{n,s}) \geqslant n - 1 + \left[\frac{7s + 2}{4}\right] = n + 1 + \left[\frac{7s - 6}{4}\right]$, which is the second part of (35) for n.

If $s = 1$, then we have $\pi(c_{n,1}) = -1 + \pi(\Sigma_{\sigma=1}^{\infty} b_{n,\sigma} a_{4\sigma,1+\sigma})$. A calculation shows that the coefficient of g^2 in S_4 is $82 \cdot 7^2$, that is, $a_{4,2} = 82 \cdot 7^2$. This, together with the first part of (34), implies that $\pi(b_{n,1} a_{4,2}) = \pi(b_{n,1}) + \pi(a_{4,2}) = n + 2$. On the other hand, by (4) and (34),

$$\pi\left(\sum_{\sigma=2}^{\infty} b_{n,\sigma} a_{4\sigma,1+\sigma}\right) \geqslant \min_{\sigma \geqslant 2} \{\pi(b_{n,\sigma}) + \pi(a_{4\sigma,1+\sigma})\}$$

$$\geqslant n + \min_{\sigma \geqslant 2} \left\{\left[\frac{7\sigma - 4}{4}\right] + \left[\frac{10 - \sigma}{4}\right]\right\}.$$

The minimum occurs for $\sigma = 2$ and we find that $\pi(\sum_{\sigma=2}^{\infty} b_{n,\sigma}a_{4\sigma,1+\sigma}) \geqslant n + 4 > n + 2$. By (4) we conclude that $\pi(c_{n,1}) = -1 + \pi(b_{n,1}a_{4,2}) = n + 1$. This proves the first part of (35) for n.

Next, assume that $n = 1$. In this case, $7c_{1,s} = 7a_{4,s+1} + 7^2 a_{8,s+2}$, so that

$$\pi(c_{1,s}) \geqslant \min\{\pi(a_{4,s+1}), 1 + \pi(a_{8,s+2})\}$$
$$\geqslant \min\left\{\left[\frac{7s+2}{4}\right], \left[\frac{7s+5}{4}\right]\right\} = \left[\frac{7s+2}{4}\right] = 2 + \left[\frac{7s-6}{4}\right].$$

If $s = 1$, then $7c_{1,1} = 7a_{4,2} + 7^2 a_{8,3}$, or $c_{1,1} = a_{4,2} + 7a_{8,3}$. Now $\pi(a_{4,2}) = 2$, and $\pi(7a_{8,3}) = 1 + \pi(a_{8,3}) \geqslant 1 + 2 = 3 > 2$. Thus, by (4), $\pi(c_{1,1}) = \min\{\pi(a_{4,2}), \pi(7a_{8,3})\} = 2$. This completes the proof of (35) in the case $n = 1$.

(c) The proof of the theorem will be complete if we can show that if (35) holds with a fixed $n \geqslant 1$, then (34) holds for $n + 1$. Assume then that (35) holds for n. By Lemmas 9(b) and 12(b),

$$\sum_{s=1}^{\infty} b_{n+1,s}g^s(x) = L_{2n+1}(x) = U\{\phi(x)L_{2n}(x)\}$$
$$= U\left\{\phi(x)\sum_{\sigma=1}^{\infty} c_{n,\sigma}g^{\sigma}(x)\right\} = \sum_{\sigma=1}^{\infty} c_{n,\sigma}U\{\phi(x)g^{\sigma}(x)\}$$
$$= \sum_{\sigma=1}^{\infty} c_{n,\sigma}\frac{S_{4\sigma+1}}{7g^{\sigma}(x)} = \sum_{\sigma=1}^{\infty}\frac{c_{n,\sigma}}{7}\sum_{p=1}^{\infty} a_{4\sigma+1,p}g^{p-\sigma}.$$

It follows that $7b_{n+1,s} = \sum_{\sigma=1}^{\infty} c_{n,\sigma}a_{4\sigma+1,\sigma+s}$, a finite sum. Hence, by (4),

$$1 + \pi(b_{n+1,s}) = \pi(7b_{n+1,s}) \geqslant \min_{\sigma \geqslant 1}\{\pi(c_{n,\sigma}) + \pi(a_{4\sigma+1,\sigma+s})\}.$$

By Lemma 11, $\pi(a_{4\sigma+1,\sigma+s}) \geqslant \left[\frac{7s-\sigma+1}{4}\right]$, so that

$$\pi(b_{n+1,s}) \geqslant n + \min_{\sigma \geqslant 1}\left\{\left[\frac{7\sigma-6}{4}\right] + \left[\frac{7s-\sigma+1}{4}\right]\right\}.$$

Once more the minimum occurs for $\sigma = 1$, and we find that $\pi(b_{n+1,s}) \geqslant n + \left[\frac{7s}{4}\right] = n + 1 + \left[\frac{7s-4}{4}\right]$. Thus the second part of (34) is valid for $n + 1$.

If $s = 1$, we have

$$\pi(b_{n+1,1}) = -1 + \pi\left(c_{n,1}a_{5,2} + \sum_{\sigma=2}^{\infty} c_{n,\sigma}a_{4\sigma+1,\sigma+1}\right).$$

By (35), $\pi(c_{n,1}) = n + 1$, and a calculation of S_5 shows that $a_{5,2} = 190 \cdot 7$, so that $\pi(a_{5,2}) = 1$. Thus $\pi(c_{n,1}a_{5,2}) = n + 2$. On the other hand, by (4) and (35),

$$\pi\left(\sum_{\sigma=2}^{\infty} c_{n,\sigma}a_{4\sigma+1,\sigma+1}\right) \geqslant \min_{\sigma \geqslant 2} \{\pi(c_{n,\sigma}) + \pi(a_{4\sigma+1,\sigma+1})\}$$

$$\geqslant n + 1 + \min_{\sigma \geqslant 2} \left\{\left[\frac{7\sigma - 6}{4}\right] + \left[\frac{8 - \sigma}{4}\right]\right\}$$

$$= n + 1 + 2 + 1 = n + 4 > n + 2.$$

Thus, by (4), $\pi(b_{n+1,1}) = -1 + \pi(c_{n,1}a_{5,2}) = n + 1$. Thus (34) holds for $n + 1$ and the proof is complete.

INDEX

INDEX